Doing English in Asia

Doing English in Asia

Global Literature and Culture

Edited by
Patricia Haseltine and Sheng-mei Ma

LEXINGTON BOOKS
Lanham • Boulder • New York • London

Published by Lexington Books
An imprint of The Rowman &Littlefield Publishing Group, Inc.
4501 Forbes Boulevard, Suite 200, Lanham, Maryland 20706
www.rowman.com

Unit A, Whitacre Mews, 26-34 Stannary Street, London SE11 4AB

British Library Cataloguing in Publication Information Available

Library of Congress Cataloging-in-Publication Data

Names: Haseltine, Patricia, editor. | Ma, Sheng-mei, editor. | Matsumoto, Kazuhito, 1952- "Exophony" in the midst of a mother tongue. Container of (work):
Title: Doing English in Asia : global literature and culture / edited by Patricia Haseltine and Sheng-mei Ma.
Description: Lanham : Lexington Books, [2016] | Includes bibliographical references and index.
Identifiers: LCCN 2016001514 (print) | LCCN 2016002752 (ebook) | ISBN 9780739192009 (cloth : alk. paper) | ISBN 9780739192016 (Electronic)
Subjects: LCSH: English language—Study and teaching (Higher)—Asia. | English language—Technical English—Study and teaching (Higher)—Asia. | English literature—Study and teaching (Higher)—Asia. | Literature and globalization—Asia.
Classification: LCC PE1068.A7 D65 2016 (print) | LCC PE1068.A7 (ebook) | DDC 428.0071/05—dc23
LC record available at http://lccn.loc.gov/2016001514

∞™ The paper used in this publication meets the minimum requirements of American National Standard for Information Sciences—Permanence of Paper for Printed Library Materials, ANSI/NISO Z39.48-1992.

Printed in the United States of America

Contents

Preface vii
Sheng-mei Ma

Introduction: Toward a Cosmopolitan Everyman: Cultural Negotiations of English Foreign Literature Teachers in East Asia 1
Patricia Haseltine

Part I: Language Politics across Asia

1 "Exophony" in the Midst of the Mother Tongue: Resources between Languages 17
Kazuhito Matsumoto

2 Unfinished English: Stories from the Other Other 29
Will P. Ortiz

3 Using Chopsticks to Read Knife 'n' Fork English: Teaching American Literature in Taiwan 53
E-chou Wu

Part II: Culture and Language Study in Taiwan

4 Teaching Early British Literature in Southern Taiwan 69
Elyssa Y. Cheng

5 Teaching English Language and Literature in the Asia-Pacific Region: Environmental and Ecocritical Contexts 87
Iris Ralph

6 Interacting with "Other" Places in Modern Poetry through Hypermedia 99
Emerald C. L. Ku and Patricia Haseltine

Part III: Drama and Ethos beyond Borders

7 Shakespeare's *Cymbeline* in the College English Classroom during the Sunflower Student Movement 121
I-Chun Wang

8 MOOC "Global/Local Shakespeare": New Approaches to Teaching Shakespeare in Taiwan and Beyond 133
Yilin Chen

9 Modern Renaissance Education in Taiwan's Departments of Foreign Languages 147
Bernard Montoneri

Index 161

About the Editors and Contributors 165

Preface

Sheng-mei Ma

The co-editors are the producers and the products of doing English in Asia, specifically, the study of English literature and culture. As today's lingua franca, literatures and cultures authored in English have become synonymous with global literature and culture, the kinship supported by a belief in the inherent values of globalization and its official tongue. The aged English language is wedded, even bond-maided, to the millennial poster child, "globalization," the buzz word that seems to end all debates. True, the dominion of English spans transnational production and consumption; its practical functions endow English speakers with enormous advantage, even social advancement. True, globalization has drastically redrawn our world and reprogrammed our minds, although no one quite knows what it is. Perhaps its power, like God's, derives from its unfathomability. In this process of popularization, indeed, fetishization, a cult-like mystique anoints globalization and its fair English maiden. While in our frenzy of flattening and surfing the globe and friending friends' friends on Facebook, cultural intricacies and in-depth investigation have given way to short-term, utilitarian concerns such as "what is the return for my investment" in speaking and writing English fluently?

Like a diamond, excellent English, or any other language in its purest form, bathes the user in a halo of splendor. Reflecting all lights, this thing of beauty, this supreme gem, stands as a reflective pool, a still center of self-reflexiveness. In our busy, business world, the "supreme" tongue can also shear through and demystify any illusion, including the previous one on an essentialized, purified language, whatever that is. Both editors have been drawn to this mysterious duality of English ever since they chanced upon each other years ago. Having earned her PhD in folklore in the United States, Patricia Haseltine, with a devoted interest in the local cultures of Taiwan, has taught English here for over thirty years. Sheng-mei Ma was her student at Tamkang University and a product of Taiwan's English education.

Their life courses diverge in that one is an American who decided to stay on this island; the other a Taiwanese who opted for life on the North American continent after his PhD Haseltine is the lynchpin of this project, mellowing the edginess of this preface, setting the tone for the collection, indefatigably recruiting and revising pieces. Indeed, she embodies doing English in Asia of the most selfless kind throughout her career, inspiring students like Ma to venture down this rarely trodden road for English majors in Taiwan.

Two different angles zoom in on the topic like a crosshair: "doing English in Asia" sharpens or blurs into "doing Asia in English," which happens to be part of the collection's intentions. We wish to write *that* Asia, surely not its entirety, intertwined with Anglo-American cultures, to ponder the power associated with the Anglo transformation. The crosshair metaphor also invokes the possibility of having "good" English creolized, done in, by localized, indigenized variants of Englishes, or having "traditional" Asia defaced, done in, by Anglophilic, Westernized extreme makeover. These scenarios of linguistic corruption and of nostalgia for the past risk essentializing, respectively, English and Asia, perpetuating the myths either of a superior foreign tongue that lies beyond reach or of a romanticized loss within oneself. Given the fluid and hybrid global village, the habitat for comfortable urban populations across the first world, such scenarios sound rigid and ideological. Yet this dismissal ignores the power dynamics driving either the mimicking of the West or the bemoaning of the lost self. The business of cashing in on the fads of English or of the golden past is obvious in English cram schools all across Asia guaranteeing success in entrance examinations and in the cottage industry of nostalgia in Asian pop culture.

Ironically, Anglicization/globalization and nativization/localization are doppelgangers shadowing each other in the twenty-first century. Because metropolitan Asia has largely remade itself in the image of the modern West, the urge to retrieve its indigenousness and authenticity grows ever stronger. Put another way, the popularity for an introverted home-coming and soul-searching in Asia is a reaction to modernity, which assuages the fear of becoming hollow men in Western suits with diamond watches and rings. Cast in Robert Louis Stevenson's terms, we wonder if it is valid to consider which is Jekyll and which Hyde. Since the two are actually one, Stevenson's ambiguity immediately rubs off on the subject at hand. Despite the progress and cachet globalization brings, as does the gentlemanly "Dr." Jekyll, it actually kills ("Je" for "I" in French). On the other hand, the beast hidden within exteriorizes as the animal hide. It is futile then to ask which is desirable and which unbecoming, if modernization and its reverse—that is, the primitive, the pre-modern, and the rural that lies at the heart of localization—are both presented as beautiful. Are they not our Janus-faced fantasy of self-realization and guilty overcompensation?

A case in point: amidst Taiwan's favoring of *yangpian* (foreign films) over *guopian* (national films), there exists an inward-looking, Taiwan-oriented counterforce in documentaries and TV programs. The twin phenomena bespeak at least two segments of the population repulsed by the other half's taste. As Taiwan constitutes one body politic, notwithstanding its ethnic, class, gender, and other divides, what is sweet as opposed to bitter rests ultimately on the same tongue. To further complicate the Taiwan case, or the Asian case, for that matter, Taiwan's aborigine cultural productions connect with the indigenous movement globally, illustrated by the series of activities from 2006 through 2010 called *Nandao dangdai yishu* (Contemporary Austronesian Art) at Kaohsiung City Museum, documented in a set of four DVDs (see http://autronesian.kmfa.gov.tw). True to its English title, visual and performing artists hail from New Zealand (Michel Tuffery, Jim Vivieaere, and Greg Semu) as well as from various tribes in Taiwan. Three New Caledonian artists (Jean-Michel Boene, Joseph Poukiou, and Jean-Jacques Poiwi) and a New Zealander Tui Hobson are included in the museum Web site but not in the DVDs. Beyond this confusion of resident artists, a self-splitting plagues its very name. While the Chinese title, *Nandao* (southern islands), is the romantic catch-all of Taiwan cultural practitioners, it bears little resemblance to its English translation of Austronesian, with the precise, geographical and anthropological roots. The lyrical "genes" of the Chinese language gravitate to the atmospheric, sultry South Seas, which jars against the science-inflected Anglo twin. Likewise, the DVDs' exclusiveness of two countries' representatives militates against the amorphous term of south-sea islands. Descriptions far thicker and analyses far more penetrating than the preceding that probe into the Taiwan-Chinese, global-local knot will enlighten us in the pieces to come.

Anchored in a year-long collaboration at Providence University in Taichung, this project is tied to Taiwan, an unstable, indeterminate sovereign state not recognized as such, cowering under Goliaths' shadows cast across the Taiwan Strait, the Pacific Ocean, and perhaps the Japan Sea. From that marginalized center, we radiate out to other parts of Asia. This reverse movement from the fringe of Taiwan begins to throw long-established power structures off balance. We have included one piece from the English-speaking former colony of the Philippines, knowing full well that this breadth and coverage of Asia may present a host of different issues. Of course, each nation has its own specific history of colonialism and the attendant Anglicization/Nipponization since the nineteenth-century. In addition, each culture has its unique characteristics, ranging from the severe aesthetical regimen and the unsettling order of Japan and Korea to the chaotic entrepreneurship and the unsettling disorder of China and Taiwan. Undoubtedly economic miracles and powerhouses each and every one of them, the human, environmental, and cultural costs may prove much too dear, evidenced in part by the pervasive sidelining of humanities—the study of language, human mind, and culture—in higher education.

Such a social reality, compounded by disparate histories and cultural traits, results in educational policies in the region that pursue modernization by means of science and technology and, of course, the language and culture of the West. It would be foolhardy to generalize so many distinct areas and nations, but postwar U.S. influence has shaped all of them, including Maoist China reacting vehemently against *xifeng* ("the wind from the West"). Postwar U.S. aid to the rubbles of Asian economies invariably prescribed the English language as a bitter pill for the good of Asian patients. Right from the outset, English was laced with business interests and human uplift, promising the spread of the American dream to Asian realities. No matter how much of a tongue-twister, body-contortionist it is, English marks the path to personal accomplishment, a neo-colonial revenant of the British Empire. This U.S. Empire contends in our new millennium with the rise of China, neo-colonialism vying with an expedient Confucianism and the "Asian" way, credited with creating the litter of mini-dragons and the one biggie. Against this background history and contemporary politics, our collection threads together Asia via one common denominator—the teaching of English literature and visual culture as global literature and culture.

Having critiqued the cult-like mystique of, à la Mikhail Bulgakov, the Master and Engarita, we clearly do not see ourselves as mindless cheerleaders of globalization; we are, to paraphrase Shelley's "A Defence of Poetry," its "unacknowledged" umpires. This very claim may elicit derision of us romantic dreamers. Yet such Yeatsian "passionate intensity," coupled with the best's "conviction," has sustained us for decades, compelling us to resist the temptation to lapse into English education per se, both in what we do professionally and, in particular, in editing this collection. The trend across Asia is to substitute Applied English (Business English, Tourism English, etc.) for English Literature. From our perspective, it would not be a hyperbole to describe the trend as a crisis, requiring concerted efforts to repel the invasion. En guarde! Here goes one riposte! There was once a joke at Sheng-mei Ma's graduate program: "If you can't write, you teach. If you can't teach, you teach people how to teach," that is, education. What started out as literature students' self-mockery as failed writers turned into biting satire against their English education rivals. That such a joke circulated among highly stressed English Department graduate students suggested they had already felt the exponential growth of education and the threat against the study of literature.

This threat has become the occupying force in non-English speaking countries treating English as a mere tool for business and trade, akin to the mercenary, Orientalist exploitation of the literary to serve the erstwhile British Empire. Acting in proxy for a greater power yet to be named, English words today are made to carry, via Internet and texting and documents, quantities of dollars and cents and numbers and dates; they have been retired from the quality and beauty of their inner core. The ideal tool to communicate facts, Applied English trivializes and cheapens as it elbows out the long tradition of art and culture. The sad fact of Asia's academe today is that professors, students, and society at large gravitate to the lowest common denominator of commodifiable knowledge of English (i.e., the four basic skills of reading, writing, listening, and speaking), without much thought of what higher purposes they are to serve. In fact, college administrators and scholars in Taiwan mindlessly reiterate business terms such as *xinxiao jihua* (marketing strategy), *wenchuang chanyie* (literary and creative industry). On the other hand, even fine writing and culture are shot through with ideologies such as Orientalism and colonialism with respect to English in Asia, from missionaries holding the King James' Bible or the Modern English Version to contemporary writers like David Mitchell indulging in exoticized, eroticized Asianness. But only keen eyes trained in critical thinking can analyze and refute these ideologies. The same parsing can be applied to advertising, marketing, business, new media, or capitalist flow in general. Cultural critics are indeed the conscience of global capitalism—its excesses and idiocies as well as its triumphs. To launch insightful critiques, the English language and its cultural history must be brought to bear on the subject at hand, a critical process requiring extensive immersion in literature and culture.

Many of the contributors are friends with either of the co-editors. They are a combination of established and early career scholars, representing the tradition of English education in Asia and its future. In the editing stage, this collection, unabashedly, unapologetically, builds on personal bonding and camaraderie, a positive spin of the Asian concept of *guanxi* (connections). First of all, collections in Western academe, all too often, run on the same principle, except that editors disingenuously disavow it. By contrast, we embrace it as the perfect way to channel our common interests toward a precise goal: how to not only preserve the study of literature but also to broaden it to cultural studies by means of critical analysis in the English language. In the publishing stage, the book proposal was submitted to a number of presses without any personal connection, in accordance with the typical and somewhat idealized Western academic model of blind submissions and peer reviews. To accomplish this, contributors are urged to draw from their years of experience teaching English in Asia, publishing in their own fields, and ruminating on the course of English. All the pieces blend professional expertise as readers of texts with personal reflections on the dark night of the soul, wrestling with the stranger by the Pacific Rim. That stranger is slippery indeed and continues to morph throughout the long match. Its "good" English slurs into various accents and Asianized mindsets to the extent that both editors (and perhaps others) begin to hear themselves in it. Cultural hybridity has truly enriched us even as we devote ourselves to teaching *English* Literature.

Contributors reflect on what works and what does not work in the context of Asia, what they do to navigate and survive in Asian universities, what triumphs and tribulations they have experienced. How have they reshaped their own writings and pedagogy in the face of non-native speaking students with an occasional flicker of fire in their eyes and in the face of a myopic bureaucracy fixated on ranking and SSCI-indexed journals? Their research must have been affected by administrative and professional demands. What are the critical insights into texts as a result? How do they negotiate the troubled waters on this side of the Pacific? As each contributor brings unique insights from his or her own positionality and experiences, it is of particular interest to see how they in their profound wisdom address the tension between the personal, experiential pull of the title "Doing English in Asia," and the universal, humanistic pull of the subtitle, "Global Literature and Culture." Inherent contradictions abound in the book title. Rather than the formal, dignified "Teaching English," we elect a vernacular "Doing English" to reflect practical, commercialized, and job-related pressures on higher education—not so high after all. Rather than confined to an ivory tower onto itself, "Doing English" is intellectual labor that falls within the context of non-English speaking Asia, which in turns falls within the even larger context of global culture. The old-fashioned, idealistic long-term pursuit of ideas and humanity contrasts sharply with the urgency of practical applications and quantifiable results in the short term. With the common tool of English, how do the contributors turn it into a transcultural, translational methodology ideal for the twenty-first century split between these opposite pulls?

We are well aware that calling attention to globalization's reigning mystique may raise some eyebrows, but a certain amount of edginess, even shock, is to be expected whenever we yoke together such alleged incongruities as East and West. As editors, how do we relate teaching English to non-native speakers there to publishing this collection for, primarily, native speakers here, especially when "we here" could care less about how they do "us" there? This issue of relevance motivates the very question of how English education in Asia connects to globalization, and to publishers and readers outside of Asia if this collection is to see the light of day. Specifically, the editors wonder how to do literature in an Asian job market favoring Applied English or ESP (English for Specific Purposes, or, pardon the expression, English for the Simple Person). This problem of relevance reveals the age-old power dynamics with a new twist: the fringe (of literature and culture) must woo the center (of business and commerce), not the other way around. The editorial challenge leads to contributors' dilemma. Juggling the two balls of our past and our future is familiar to these contributors who have spent their entire careers in a carnival called higher education. In a circle, the past of having done English changes hand into the future of doing the globe, which instantly vanishes into the air. Knowing these friends, we are not surprised to find pieces talking back in dissenting views, performing new aerial stunts with their magical English words. In contributors' various "case studies," we respect their individual style and vision, striving to preserve their distinct voices, even if some happen to jar against American sensibility and academic protocol. The collection, after all, is *Doing English in Asia*, not in the United States.

Originally, Sheng-mei Ma commissioned a piece on the "dark side" of Taiwan's English teaching, namely, evening cram schools (*buxiban*) that few college professors, Ma included, have ever had the interest of venturing into. *Buxiban* means, literally, supplementary/alimentary learning class/institute, with the positive connotation of "*bu*," nutritious and health-enhancing supplements in Chinese herbal medicine. Indeed, to get the whole picture of doing English, one needs to look at this phenomenon all across non-English speaking Asia, this booming industry catering to high school and college entrance examinations as well as job-related English-language proficiency tests. The Taiwanese returned graduate student—say, an X—from whom Ma commissioned the piece has taught English in such cram schools for over a decade, the income having enabled X to attend the PhD program. X submitted a draft that defended the practice of cram schools and protested their lowly status. After having received extensive, detailed comments on style, argument, and cram school specifics—student enrollments, costs, placements, test score improvements, curricula, identity conflicts, credential problems, and academic stigma—X decided to meet with the cram school director for such data and, as it turned out, for blessing. The director thought publishing the piece "would not be proper" and X withdrew, which must have been a painful decision. It also pained Ma that despite the humiliation of having been derided by one of X's own professors as *heipai jiaoshi* (black-labeled, uncertified, second-rate, and fake instructor toiling outside academia), X preferred not to talk back. Reminiscent of the double invocations that conclude Melville's "Bartleby, the Scrivener: A Story of Wall Street" (1853), this graduate student X's silence speaks for the humanity of Taiwan's underclass teachers walled in by cram schools and social norms.

If only Ma had been a graduate student fortunate enough to have his professor commissioned a piece, he would have jumped at the opportunity of breaking into the big league, the global marketplace of ideas, and would have voluntarily gone through draft after draft, if need be. But X's priorities were apparent and understandable: it was safer to disappoint Ma than to offend the cram school director, who is likely to be the boss for years to come, given Taiwan's dismal job market for an oversupply of "home-grown" PhDs. As such, what is not in *Doing English in Asia* is as important as what is, the darkness of unnamed X-es as illuminating as the light shed by contributors. Among the glaring lacunae is Asia's evening cram schools, the parallel track to English education in regular schools. Rarely analyzed in academic circles, these shadow schools and classes are thriving; they meet the demand for Applied English and English-language test-taking, "Instant Englishes" like instant noodles to satiate the hunger of the people. Once the mainstay of the humanities, the joy of slow food that simmers in university's Anglo-American and global literature courses has increasingly been perceived as elitist, epicurean. Such joy has been replaced by corporatized, profit-driven, cram school–style fast food. How do we do English in Asia for the new millennium—the fine dining we can no longer afford or the cheap empty calories that make us feel so good, so filling? Both? One thing is for sure: whichever way we choose to do English, we do it for ourselves.

Introduction

Toward a Cosmopolitan Everyman: Cultural Negotiations of English Literature Teachers in East Asia

Patricia Haseltine

For those who teach English in Asian countries, a major challenge is how to translate between cultures and construct a bridge to communicate about the things and ideas in at least two cultures or subcultures. In the East Asian sector of the global arena, English shares the historical moment with the dominant languages of Taiwan, China, Japan, Korea, and the Philippines, and it ubiquitously resounds in varied contexts, from the practical applications in business relations and the popular cultural scenes of social media to refined academic conferences. Such a situation requires that young people wanting to distinguish themselves must take on the arduous task of learning English. For me, growing up in the United States, learning the foreign languages of first Spanish, then German and Chinese was a free choice I made in accordance with one of my native-born intelligences. However, for many students learning English in my classes in Taiwan, foreign language study looms large as a contingency of their geopolitical environment and history.

The significance of learning English in East Asia increased considerably in the twentieth century as functionality took hold in earnest following the end of World War II. In 1910, the Korean Empire was annexed by Imperial Japan and only regained its autonomy as a nation in 1945. The Philippines had been a colony of Spain from the 1500s, but after the Spanish-American War, it was ceded to the United States in 1898. A struggle for independence ensued, culminating in the drafting of a constitution in 1934. However, without full independence being achieved, the Philippines were subjected to Japanese imperial rule from 1942 until 1945. Universities were in turn established under Spanish, American, and Japanese rule.

Taiwan remained a colony of Japan from 1895 to 1945, and the Japanese language was taught on all levels of education. In 1949, when American troops bolstered the Chiang Kai-shek regime in its Taiwan refuge from Communist rule over mainland China, English served a special military purpose. But after the withdrawal of American troops, Taiwan became more culturally insular, and English remained important for the purpose of training excellent students, some of whom would go abroad and return to stand in the ranks of the higher echelon of Taiwan academe and officialdom. At that time, English was also a purveyor of aesthetic appreciation for Western "developed" culture. Translation methods of language instruction prevailed, and literature classes dominated the English curriculum.

Once useful to those who catered to the needs of the army personnel and their families, then cultivated in the exotic hothouse of the universities as the language of university literati, English is now touted in many Asian countries with a global profile as something for Everyman, at least for every businesswoman and consumer. It is a commodity item itself in the now highly prosperous cities. As in the medieval allegory of the Seven Deadly Sins, this possession lies in chests and hoards—vocabulary hoards, and its amount is recounted periodically by proficiency exams, both foreign and home-grown. Concretely representing the sin of avarice, the attainment of goods is a stronger impulse for learning English than any other cultural requisite. And the person who counts English among her possessions is not likely to repent like Everyman did at the end of his life, even though she may be wrought by spells of depression when business doesn't run smoothly. The goal of working for a multinational business makes acquisition of the language a stronger impulse for learning language than appreciation of the language for its own sake or for the sake of sharing universal human feelings or cultural integration. In such an environment, to profane an image from W. B. Yeats, the once humanistic-inclined teacher is made to feel like a mechanical nightingale of hammered gold singing to keep drowsy would-be chemists, entrepreneurs, and their future secretaries awake.

Yet there is some natural light penetrating this glittering darkness of functionality and mammonism. It is a light reflecting across special areas of cultural negotiation designed by literature teachers, such as those represented in this collection. Because of the high stakes of students' acquiring the language, English teachers, domestic and foreign, are always being watched and watching themselves, but the assiduous teacher can discover devious ways to escape the language-only-for-practical-purposes trap. These teachers are also justifiably wary of making the English literary canon a hegemonic tool. As an alternative, eschewing universalism, many treat literature as a way to explore similarities and negotiate differences between cultures. This book attempts to share some of these attempts, if not yet methods, of cultural translation. As he sees how texts from different cultures overlap and are rewritten or reinterpreted subjectively from different cultural viewpoints, Everyman is invited to assume a distinctive multidimensional role in his own intercultural life.

The literary sites where cultural texts meet and interact with one another in ways that evoke cultural translation and negotiation are numerous. In addition to new approaches to the literature canon, several distinct areas have been wrought into multiple perspectives in Asian university classes of English. They would include children's literature and local and indigenous literatures, both of which have the power to change cultural concepts. Also having an impact are literature in film courses and Asian culture courses conducted in English that have the power to turn the insider's culture outward and the outsider's culture inward.

The papers collected in this volume show how Asian teachers approach the English literature canon from their own multiple perspectives and how texts are domesticated in Taiwan, the Philippines, and Japan. Although the writers have varied identities, teaching personae and research focuses, the reader will find that in their "doing English" they are not making English solely a commodity. The functionality serves higher purposes than personal gain: that of promoting ecological concern for the environment, of understanding and comparing classical literature and cultures, of explaining the capitulation to ancient and modern imperialisms, and other philosophical considerations. From their varied insights, it may be surmised that teaching English in Asia in our historical moment is steering toward new trajectories.

INTERCULTURAL NEGOTIATIONS

To find one's equilibrium in another nation's living culture, negotiations of various kinds are indeed necessary. Over the years, foreigners teaching English in Asian universities and training schools have had the distinct social role of "native-speaker" to play. And no matter how long they have stayed, they have usually remained just "foreigners teaching English" in the perceptions of most local people in spite of their own multidimensional cultural emergence and sometimes near-native communicative ability in local languages. It is their main function or employment that marks them, that of being English teachers. Like a chess game, the lines are marked and one's intrusion into the foreign territory may be tolerated only by making designated moves. Or, as in the traditional Chinese game of Go (圍棋), cultural territory is parceled out by moves from each side, and there is an ever present chance of being surrounded, captured, and told to return home.

The complex human relationship of "face" in most East Asian cultures is another important aspect of foreign teachers' experience. It may seem to observers that foreign teachers are accorded "special privileges" manifested in communities where politeness is prioritized. With experience, one finds such privileges are illusory. In faculty meetings, for instance, often decisions are made before meetings in order for consensus to be achieved in the meetings themselves. Trying to be innovative to achieve improvements, a foreign teacher's spontaneous suggestion may be politely heard out, only for the innovative idea to be subsequently ignored as irrelevant.

Different cultures have their different codes of behavior for various contexts, and the foreign language departments often seem to be anomalous entities jostling between concepts from the native and the foreign culture. Some participants are playing what seems to be a Western aggressive, individualistic chess game while others are following strict hierarchical rules of behavior. The locally born English teachers, those originally hailing from the Asian country, savvy to Western codes, ostensibly can more easily play two roles at once—insider and outsider, teaching the foreign literature but switching to cultural insider when expedient.

No matter whether of the home-grown variety, well-traveled or not, or of the increasingly more exotic one, the English literature teacher herself resists being a commodity. Engagement in problem-centered intercultural dialogue has thus evolved as a way to counter the narrow focus on language study in literature teaching. Most of the negotiated topics and themes derive from the context of the cultural wars in Western academia of the 1960s and 1970s. The waves have been rippling from the United States and Europe over the years into the various English departments abroad. Our currently committed concern in this book, *Doing English*, is in fact a response to that ripple in our many various ways, and includes counter-waves and crosswinds. That is, communication on special themes and genres includes trans-Pacific, as well as inter-Asian, interchanges.

Walter Benjamin has provided an ideal view of inter-language communication with the idea of a transcendently pure understanding that crosses linguistic and cultural borderlines. Whether or not one accepts such an idealism, those who "do" English through literature in foreign countries make particular choices consciously or subliminally about their intercultural communication. One of these choices concerns the question of universalism. Kwame Anthony Appiah, in addressing beliefs and values, makes an important distinction between universalists and cosmopolitans: "Cosmopolitans suppose that all cultures have enough overlap in their vocabulary of values to begin a conversation. But they don't suppose, like some universalists, that we could all come to agreement if only we had the same vocabulary" (57). I would like to add that the cosmopolitan teacher recognizes how language is deflected or refracted by different kinds of language use. She works with these differences.

Engagement with literary texts yields a space for freedom to entertain connotations of language and semiosis in different contexts. However, foreign language students rarely make the connection between their reading and their openness to other cultures. When I asked foreign language students[1] in Taiwan whether they consider themselves cosmopolitan, a good number said no, but they are "trying to be"; while others cited their learning English or Japanese as making them cosmopolitan. Being more open-minded is seen as a goal. Side-stepping the importance of their language-learning and reading of foreign literature in school, many claim that it is their use of the Internet and their travel that makes them global citizens. One said, I am not cosmopolitan because I am shy. It could be surmised that it is better, perhaps socially safer, to remain humble about one's acquisition of cosmopolitanism.

As Jessica Berman points out, the meaning of the word cosmopolitan has suffered from pejorative connotations, especially in the nineteenth century when it could indicate the lack of roots and even vagrancy. In the positive sense, however, it is a post-Enlightenment humanism searching for harmony in the world, all the while supporting national economic and cultural interests (31–36). Related to this different perception of the value of being cosmopolitan, twentieth-century teachers make decisions opting to maintain literature's "foreignness" or produce a familiarizing "domestication."[2] That is, as in language translation, interpretations of foreign language texts can veer toward emphasizing the source culture's ideas and expressions or toward accommodating the meanings and interpretations from the target culture. In interpretations of, let's say, William Wordsworth or Henry James, some teachers require students to immerse themselves in the words and ideas of the foreign author, while others let the students keep their local culture in mind and allow a wider range of interpretation based on Asian sources.

Henry James's *Daisy Miller: A Study* (1879) is a work that brings out problems of intercultural understanding and the development of cosmopolitan identity in a simple manner for students to contemplate. In the story, James uses a pun on the word "parts" in a way that appeals to the imagination of the reader. While Daisy herself realizes that she is "not of these parts" when she is positioning herself as a foreigner, her disgruntled, more cosmopolitan paramour later states that he has "lived too long in foreign parts" and failed to appreciate Daisy's American naïveté (50). The play on the word "parts" is obvious when undergraduate students interpret its meaning in various ways. Although "parts" bears predominantly a geographical context in the story, referring to foreign countries or places distant from America, under the student's readings "parts" is also a word contrasting with a "whole." Winterbourne has a divided identity as he is composed of both native and foreign parts to his social personality. He could also be considered a "man of many parts." In Renaissance usage, this would mean someone with many talents, a polymath, a cosmopolitan intellectual role discussed by Bernard Montoneri in this book.

Both Daisy and Winterbourne are characters that mirror the cultural dilemma foreign language students face. Partial assimilation to a foreign culture alienates one part of the self from another one. Cultural hybridity has an emotional aspect as one trans-valuates one's own cultural constraints and conventions and tries on new ones. Brian Lee comments on how in James's *The Portrait of a Lady*, Isabel has the "determination to see, to try and to know." She is attracted to European civilization, "yet her moral integrity guarantees her immunity" (36). As students become more cosmopolitan, a certain degree of double-ness, if not to the degree of cultural hybridity, is to be expected.

As an English teacher in Asia works along with students to fathom the many meanings of human situations in literary texts, the overlay of language, culture-meeting-culture, becomes an aesthetic site in the never-ending negotiation of meaning. But the teacher must be open to the students' own interpretations. The listeners' or readers' interpretation, called a "counter translation" by Jacques Rancière (69), depends upon a stance of equality between text and any attentive interpreter of the text, irrespective of his ethnic or class origins. Before discussing examples of such intercultural approaches to language teaching with a literature component, in the following I will give a brief historical sketch of the study of English literature in postwar East Asia with a special focus upon Taiwan.

ENGLISH LITERATURE STUDY IN ASIA: THE EXAMPLE OF TAIWAN[3]

In Taiwan in the 1970s, New Criticism was the dominant theory in use in literary courses in what were ironically called Foreign Literature Departments even though the only or main language taught was English. With New Criticism, one could read canonical texts thoroughly and discuss at length how discreet words and phrases in the English language were being used. This was the norm, for example, spread by the Taiwanese teachers returning from education abroad. Admittedly, close reading and translation fit nicely into the "using literature to learn language" model of teaching. Some foreign language literature teachers still use it today. The only problem is that it is an autocratic approach primarily working with the denotative level of the text, and the interpretations were more or less given by the authorities or the teachers as the masters of language.[4] Then when subjectivity arrived toward the end of the decade with phenomenology and the reader's response theories, a new breed of scholar was born in Taiwan, one that privileged subjectivity. This could not spill over into the language classes, however, since students expected definite answers as to meanings of sentences in English texts, answers they could use on objective tests. Widening one's vocabulary through reading literature was noticeably still the aim of many serious students even when reader's response theories began to win the day among the circle of literature teachers. In language classes, subjectivity of language interpretation in the classrooms remained dormant.

The confrontation between the claimed objectivity of New Criticism and subjectivity began to raise ire in the late 1970s when international conferences highlighted the criticism of Jacques Derrida. At a conference of the Comparative Literature Association of the Republic of China at Tamkang University in Taiwan in 1979 there was a particularly stormy session when a deconstruction paper fueled political innuendos concerning the rule of one political party, the Kuomintang. As deconstruction claimed the podiums at other venues, other voices also began to be heard: the English literature curriculum was expanded to include African American literature and works written by women. A favorite poem for teaching undergraduate students in Taiwan became Langston Hughes's "A Dream Deferred" that was used to connote the dreams local Taiwanese speakers had of political equality.

Concurrently, in Chinese literature courses, local-color literature with Taiwanese cultural themes by such authors as Huang Chun-Ming was being introduced in university courses. His 1969 "Gong" (鑼), as well as "A Flower in the Raining Night," (看海的日子) and "Sayonara/Zaijian" (莎喲娜啦．再見), popular short stories from the 1970s, were translated into English by Howard Goldblatt in 1980. Huang's works that attempted to render the Minnan dialect into Chinese were compared to those of Mark Twain and William Faulkner in their regional focus on downtrodden, illiterate characters. As this Nativist literature entered the universities, other voices began to be heard. However, the political dreams were still to be deferred, especially when the Mei-li Island uprising of political dissidents in the southern city of Kaohsiung was quickly quelled. And, once again, freedom of literary interpretation went underground.

The Awakening Foundation, a feminist movement begun by Lee Yuan-Chen, supported by women from various walks of life and university humanities departments, challenged the patriarchal status quo. It was a group "independent of the labor and democracy movements" (Chang 160), and the members' work for gender equity culminated in the Gender Employment Equality Law. Lu Hsiu-lien, an early supporter of the group, became vice president of the Republic of Taiwan when the Democratic Progressive Party won the election in 2000.

Coinciding with the democratization of the country, the commodification of English and its specialization into the categories of literature, linguistics and teaching English as a foreign second language was to constitute a phase lasting throughout the 1980s and 1990s. Taiwanese and Hakka, not just Mandarin Chinese, were accepted in the schools as they became more multicultural. Taiwan attempted to break out of its political insularity. Relations with other nations and global trade agreements were made. A number of countries accepted Taiwan as a state in its own right, but this reaching out to the world was determinedly countered by China's claims of sovereignty. The traditional mode of teaching English with the literary canon broke down in earnest at this time.

Although not widely developed in undergraduate literature programs, a culturally integrative approach with texts from Asian American writers worked its way into research and graduate programs. The question about whether marginal literatures spoke more closely to the condition of the non-native speaker was asked. When an English language learner in Taiwan or Japan reads a work of Asian American literature, the characters' moments of disappointment and chagrin for not fitting in to the mainstream society and the desire for seeking justice for one's own voice and ethnicity find a resonating chord. With the prismatic relations of language communication ricocheting around from culture to culture when students read works of American literature replete with Chinese traditions and ethnicity, such as *The Woman Warrior* by Maxine Hong Kingston or *The Joy Luck Club* by Amy Tan, engagement and negotiation of meanings intensifies. My co-editor, Ma Sheng-Mei, in *The Deathly Embrace* has critiqued this literature and found that the authors' self-orientalization gave them a wide readership in the United States. Granted, readers in Asian universities were not an intended readership for the sentiments of these novels, but, at least, there is a successful transmission of the quandaries of translation that arise in bicultural communication.

Is the minority stance in a mainstream culture similar to the stance of the foreign language learner? Well, this is the very bone of contention in the fear of the influence of English around the world. Minority status abroad is often deemed to be less prestigious than the position of the foreign national. That is a reason why the practice of translation is preferred to the global use of English in maintaining equal status between countries with divergent languages. At the end of the twentieth century, the change to minority literatures that would put English into a largely non-canonical post-colonial setting was countered by a rather politically conservative thrust toward criticizing nonstandard English and the desire by some teachers to reinstate the canonical authors as reading material. This move was accompanied by the cacophonic litany that the students did not have a good enough English background to do literary study. Such a rejection of literature was a double-edged sword, dismissing the non-canonical minority literatures that might resonate with ethnicity links of students, as well as canonical literature, in favor of a language-only approach. The result was to place learners in a more colonial, rather servile, relationship to the countries where English was the dominant language. It alienated students even more from the deeper currents and counter-currents of Western cultures.

It was in such a multicultural literary context and a multi-party political context that the importance of teaching English literature declined. English became just a foreign language to be learned for global communication purposes with nations that were like Taiwan, not native English-speaking themselves. Moreover, because of its perceived common value for people in all walks of life, English should be introduced to children. The cram institutions, which had formerly prepared students for study abroad and for university entrance, now expanded their clientele to include children who would use English for lower-level entrance exams. Since the cram schools, influenced by TESOL (Teaching English as a Second Language) programs designed for immigrants to the United States from around the world, seemed to have useful methods of teaching the four basic skills of English language to large numbers of students, the official response of the public school system was to use such techniques in the elementary schools. Foreigners and young Asian Americans were also enlisted to teach in summer programs or camps sponsored by Taiwan's Ministry of Education.

In general, the advice from linguists was accepted that starting to learn a foreign language in childhood is effective. One of the joys of the beginning stages of learning language is the childlike way students are asked to repeat words and phrases. The child in us all learns from games. Trying to make up for lost time for students already in the upper grades, college instructors chose books containing predominantly oral methods. Even though the students were young adults or even oldsters, a classroom climate of infantile learning and imitation arose in the upper levels, where playing language games and acting out dialogues was a technique adopted to make the experience more pleasurable than before. The only problem with this was that, lagging behind the enhanced language abilities of the next generation of students, the teaching of easy, but pleasurable, language exercises in college substituted for more advanced work in reading. Some students could already read at a higher level but would rather not spend the time and energy with the more challenging reading in English that literary works would entail. Moreover, proficiency exams to enter college in Taiwan, as in Japan and Korea, had burned them out in high school. Entering college became a reactionary time for *flaneur*-ship through the glitzy department stores of the cities of Osaka and Tokyo, Taipei and Taichung, Seoul or Manila, not a time for reading Mark Twain or Shakespeare.

On a more positive note, the attention on the child in Taiwan yielded a fortunate outcome when publishers started importing books of children's literature, and cultures began to meet in aesthetic negotiations in the field of children's literature in Asian universities. The democratic idea that children's language, the mother tongue, held a philosophy of truth and simplicity in relation to other more abstruse philosophies was a concept put forward in annually held children's literature conferences at Providence University beginning in 1997. Under the guidance of Chao Tien Yi, a philosophy professor originally from National Taiwan University and later from Providence University, the conferences were international in scope. Papers on Chinese, Taiwanese, Japanese, Korean, and English works of children's literature were delivered to an educated audience of writers, teachers and publishers from all levels and from various countries. These and other conferences in Asia reflected both the desire and the fears of intercultural children's language learning and debated the use of literature in native and foreign languages. Identities were kaleidoscopically both defined and blurred. Most participants shared the romantic concept expressed in William Wordsworth's "Ode: Intimations of Immortality from Recollections of Early Childhood" that, through the voice of the child, we all could be closer to nature. The accompanying defense of inductive education, such as that of Rousseau's *Émile*, countered the hegemonic structure of Chinese education and systemization.

In a globalism context, however, the concept of "the child" hits the connotative fan. In Asia, some of the connotations become particularly jarring, as in the dialectical opposition between nineteenth-century British Empire and the peoples of the less-"developed" world (also conceived as the North/South divide), the world's children, a concept supported by scholars such as Edward Clodd. Overlaying this background palimpsest text is another cultural text germane to our students that uses children's literature to take a subversive stance against mainstream Chinese culture. So, in general, children's literature could also be used to counter the oppressions minority cultures, Minnan, Hakka, and indigenous, within Taiwan felt. In Taiwan, the child could be seen as the oppressed Minnan local culture as opposed to the Chinese literary heritage as a cultural standard. Fear of the mainland Chinese incursions on their autonomy added to the paradox of the numerical majority of Taiwanese feeling that they had minority status. I would suggest that, when one starts to negotiate the meanings of images in the palimpsest of literary texts at the interface between cultures through English, one uncovers a learning that the Everyman could profit from in rethinking identity and addressing historical residues in global trade relationships.

Besides the multifaceted study of children's literature in transnational and interdisciplinary conferences of academia, the field of indigenous literature connects people in various departments. The study of Native American literature in English departments metamorphosed into the study of trans-indigenous literatures since Taiwan itself has a rich indigenous heritage. New literacy in Chinese led to an emerging literature by educated, imaginative individuals from some of the tribes in Taiwan. Syaman Rapongan, a graduate of a French department, articulates his relation to the Dawu (Tao) heritage of Orchid Island while Nequo Soqluman, a Bunun writer, who has a degree in ecology from Providence University, uses fantasy to rewrite the legends of Jade Mountain area. And the Puyuma poet Badai turns history and legend into modern poetry. In the last five years, there is thus a growing awareness of the literary contributions of an indigenous diaspora reaching across the ocean to Australia, the United States, and Canada. Mostly conducted by teachers of English, scholarly research on ecopoetics, like that of Huang Hsinya, a comparative literature scholar from Sun Yat Sen University, has focused on the common ecological concerns in these emerging literatures of the Pacific Region. Survival from the dangers of climate change for the islands of the Pacific requires more intercultural understanding of how to protect the ocean, and literature can provide a needed defense on this issue. For example, students reading English see that the problems faced by the Pacific islanders in Linda Hogan's *People of the Whale*, who struggle with the big business of Japanese whalers and quarrel within their own indigenous community over traditions that preserve their way of life, have similar problems to those of the indigenous peoples in their own country. The Dawu of Orchid Island off the coast of eastern Taiwan try to maintain their culture while enduring the long-term existence of nuclear waste deposited on their southern coast. And the higher ocean levels caused by climate change will eventually affect all the coastal peoples of the Northwestern United States, as well as those in Japan, Taiwan, and the Philippines. Literature articulates the common cause of survival. The future Everywoman/man engaged in business or science or humanities, simply put, shares this common cause.

TURNING THE INSIDE OUT AND THE OUTSIDE IN

When one is teaching a language, no matter which one it may be, that source language is or becomes a dominating one. It is what the students are to learn or to "do." To mitigate the dominance in the classroom, one of the most effective ways is to switch cultural positions and texts so that English is used to introduce students' own culture and literature to the teacher, either foreign or native, who can act as an outsider to Korean, Japanese, Philippine, or Chinese and Taiwanese culture, actually performing the ignorant schoolmaster role. The students are using English, but assuming the teaching role of cultural insider. In these situations, as students articulate facets of their own culture in English in cultural translation, they discuss and evaluate their own traditions. In this way, what could very well become a block to mutual understanding is overcome in the crucial intercultural contact. By using English to discuss classical literature or works from contemporary Asian writers and encountering the translations of these works, students increase their ability to articulate their own culture, sometimes with a more critical eye, seeing themselves as others might see them and weighing their own choices in life against following the traditions.

Another way cultural translation can be brought to the foreground in the classroom is to have students make aesthetic judgments about transnational films. The film director Ang Lee is from Taiwan, but he attains a multicultural reach in his themes, his selection of actors, and production. For example, the transnational films *Sense and Sensibility* (1995), *Brokeback Mountain* (2005), and more recently *The Life of Pi* (2012) are themselves aesthetic negotiations. These are Taiwanese mythopoetic interpretations of England, Western America, and Oceana. His version of Jane Austen's novel *Sense and Sensibility* was a very popular film in Asian circles. It seemed to fulfill the ideal for gender relationships among students of the 1980s and early 1990s. For them, the education in English and British culture is given an aesthetic treatment: the pace of conversation and interaction in Ang Lee's film is considerably slower than other versions and allows for the reticence young people in Taiwan used to manifest in gender relationships. With *Brokeback Mountain*, also a film of slow-paced intimacy, students imbibed the distant beautiful mountain scenery, the broken family relationships, and the homosexuality of the film. Alienation and veneration of the culture of the cowboy image was enhanced, just as the third space subversion of the mystique of American/Western white masculinity was attained in the tender emotional attachment of the cowboys in the film. For English learners in turn-of-the-century Taiwan, Ang Lee's films were cultural translations with foreign projections or reflections of themselves in a foreign milieu. Just as they were transgressions of the norms of Taiwanese relationships, they were subversions of earlier concepts of what England and America were like.

Foreign-language students who are in tune with literature see another side of the culture they study. The superficial glitz of the Western country is not so overwhelming to their sensibilities. Realities of poverty, conflict, and alienation are not only suffered by the people students know in their own local regions. Nor are these realities remote statistics or media reports. John Okada's *No No Boy*, reflecting the World War II internment of American citizens of Japanese descent, and Lawrence Yep's *Child of the Owl*, a story about a Chinese American tomboy named Casey in San Francisco's Chinatown, are bridges to students' own feelings about their local traditions and historical relations with other countries. Just as creative writers and filmmakers exercise a freedom to trope and revise cultural images, students themselves provide alternative or counter translations of what they read. Traditions may be rewritten/reinscribed, critiqued, and even parodied almost with impunity. By translating between cultures, language teachers, although they may be compromising the traditions reflected in the texts, are negotiating for old/new ground for their communities.

The commonality of our approaches in this book, *Doing English in Asia*, is therefore in our shared engagement in negotiating cognitive and cultural differences uncovered in our encounter with literatures in the study of English as a global language. The first section "Language Politics across Asia" contains pieces from Japan, the Philippines, and Taiwan that examine theoretical issues. Kazuhito Matsumoto in "'Exophony' in the Midst of a Mother Tongue" returns "exophony" to its root of "being out of familiar sound" and explores the cognitive experience of an "in-between" realm where the native and non-native languages one has learned are in a state of "Self-Forming Formlessness." Matsumoto further destabilizes English by applying the notion of exophony to any national language. Will P. Ortiz's "Unfinished English" investigates her love-hate relationship with English, being herself "a product of a colonially designed education." Launching her critique from a subversive, postcolonial positionality of "the Other Other," Ortiz discovers that English "has become a 'felt' language, surprisingly speaking to her being." E-chou Wu deploys eating utensils as metaphors for Anglo-American literary studies in "Using Chopsticks to Read Knife 'n' Fork English." He analyzes how American literary texts "have been interpreted in multi-faceted ways in a Taiwanese context." He offers insight into four different approaches to tackle American literature in the Asian classroom: functionalist, affective, parallel, and allegorical.

In the second section "Culture and Language Study in Taiwan," three geographically and ecologically informed pieces examine the role of place in the consciousness of English learners from Taiwan. Elyssa Y. Cheng's "Teaching Early British Literature in Southern Taiwan" notes the uneven distribution of resources in Taiwan—north and south as well as national and private institutions—and the resultant discrepancy of English education on Taiwan. To deal with this lack, she revises her traditional teaching pedagogy for medieval literature in favor of a "more active paradigm," which includes artistic activities and comparative videographic analysis. Iris Ralph in "Teaching English Language and Literature in the Asia-Pacific Region" illustrates basic ecocritical concepts with examples from contemporary Australian poets and hopes to inspire students to focus on similar ecological issues relevant to the Taiwan bioregion. Emerald Ku and I in "Interacting with 'Other' Places in Modern Poetry through Hypermedia" examine how Taiwan students can travel virtually to remote places represented in modern poetry through hypermedia technology. Both the rewarding, as well as the misleading, links on such a journey are explored for a New England locale in a poem by Robert Lowell and for Brazil in a companion piece by Elizabeth Bishop. Poems are taken to be "a field composed of references and representations of places and people that can be supported by hypertext links or searches on the internet."

In the third and last section "Drama and Ethos beyond Borders," ideas and images from the canonical works of the English and European Renaissance find a place in Asia's universities alongside current student interests and gender recognitions, their political aims and business prospects, and their ethical deliberations. I-Chun Wang's "Shakespeare's *Cymbeline* in the College English Classroom during the Sunflower Movement" interweaves the teaching of Shakespeare's play with the border politics and national identity roiling up during the 2014 Sunflower student protest movement in Taipei. In particular, Wang records how the political dilemma of the English trying to negotiate a viable position in relationship to Imperial Rome was interpreted from the perspective of the students' own political stance on ties between Taiwan and Mainland China. In "MOOC Global/Local Shakespeare," Yilin Chen adopts an intercultural perspective in her Shakespeare classes and online courses. She emphasizes collaborative activities and student participation, while drawing extensively from manga and Taiwanese Shakespearean productions housed in the Taiwan Shakespeare Database. Bernard Montoneri in "Modern Renaissance Education in Taiwan's Departments of Foreign Languages" first takes stock of the crisis of under-enrollment in Taiwan's institutions of higher education. He then proceeds to advocate a broad-based humanist curriculum associating Renaissance values with Asian philosophies and polymath eclecticism with modern mass media communications, such as LinkedIn and Facebook.

As teachers, the authors of this book demonstrate that the cultural values, geopolitical positions, and ecological consciousness expressed in literature all are crucial aspects of foreign language learning if one is to lead a globally interconnected life. The current turn toward a language-only curriculum may be economically practical in the short term. However, it overlooks the needs and potential of language students who have a personal stake in developing and communicating their individual and collective values cross-culturally.

NOTES

1. Students whose major was a foreign language from National Taipei University, Yuan Ze University, and Providence University in Taiwan were given questionnaires about cosmopolitanism and globalism in class in 2008, 2013, and 2015 respectively. These answers are selected from their responses.

2. These are terms discussed in the history of language translation by Lawrence Venuti in his *The Translator's Invisibility* (1995). Venuti disparages the preponderance of "domestication" in English language translations of works from other languages and believes such a practice to often be unethical "ethnocentric violence" (34).

3. This review of academic changes is based on personal observations and participation of the writer who was a member of the English Department faculties and Graduate Schools at Tamkang University, Providence University, and other universities in Taiwan, the Republic of China, in a period from 1975 to the present.

4. In a critique of post-enlightenment dominance of teacher's knowledge over learners' ignorance, Jacques Rancière in *The Ignorant Schoolmaster* (1969/1991) has defended a comprehensive approach to learning that takes seriously the students' equality with the teacher. The attentive student may learn language and other matter without being manipulated by top-down instructional methods that perpetuate inferior status.

REFERENCES

Appiah, Kwame Anthony. *Cosmopolitanism: Ethics in a World of Strangers*. New York: Norton, 2006. Print.

Balcom, John, ed. *Indigenous Writers of Taiwan: An Anthology of Stories, Essays, and Poems*. New York: Columbia University Press, 2005. Print.

Benjamin, Walter. "The Task of the Translator." *Illuminations: Essays and Reflections*. (First Pub. In German, 1955). Trans. Harry Zorn. New York: Harcourt Brace, 1968. 69–82. Print.

Berman, Jessica S. *Modernist Fiction, Cosmopolitanism and the Politics of Community*. Cambridge: Cambridge University Press, 2001. Print.

Chang, Doris. *Women's Movements in 20th Century Taiwan*. Urbana: University of Illinois Press, 2009. Print.

Clodd, Edward. *The Childhood of the World: A Simple Account of Man's Origin and Early History* (1872). New Edition. New York: Macmillan, 1917. Print.

Haseltine, Patricia L. "Too Long in Foreign Parts: An Asian Reception to Henry James's *The American* and *The Portrait of a Lady*." *IAFOR Journal of Literature & Librarianship* 1.1 (2012): 15–74.

Hogan, Linda. *People of the Whale*. New York: Norton, 2008. Print.

Huang, Chun-Ming, *The Taste of Apples*. Trans. Howard Goldblatt. New York: Columbia University Press, 2001. Print.

Huang, Hsinya. "Toward Trans-Pacific Ecopoetics: Three Indigenous Texts." Ed. Hsinya Huang and Clara Shu-Chun Chang. *Aspects of Transnational and Indigenous Cultures*. Newcastle upon Tyne: Cambridge Scholars, 2014. 113–140. Print.

James, Henry. "Daisy Miller: A Study" (1878) in *Tales of Henry James*. Ed. Christof Wegelin. New York: Norton, 1984. 3–50.

———. *The Portrait of a Lady*. (1881). London: Everyman, 1991.

Kingston, Maxine Hong. *The Woman Warrior: Memoirs of a Girlhood among Ghosts* (1976). New York: Vintage Books, 2010. Print.

Lee, Ang, dir. *Brokeback Mountain*. Focus Features, River Road, Alberta Film, Good Machine, 2005. (Based on a short story by Annie Proulx). Film.

———, dir. *Sense and Sensibility*. Columbia Pictures and Mirage Enterprises, 1996. (Based on novel by Jane Austen). Film.

———, dir. *Life of Pi*. Twentieth Century Fox and others, 2012. (Based on novel by Yann Martel). Film.

Lee, Brian. *The Novels of Henry James: A Study of Culture and Consciousness*. London: Edward Arnold, 1978. Print.

Ma, Sheng-Mei. *The Deathly Embrace: Orientalism and Asian American Identity*. Minneapolis: University of Minnesota Press, 2000. Print.

Okada, John. *No No Boy*. (1956) Seattle: University of Washington Press, 2014. Print.

Rancière, Jacques. *The Ignorant Schoolmaster: Five Lessons in Intellectual Emancipation*. (First Pub. in French, 1969).Trans. Kristen Ross, Stanford: Stanford University Press, 1991. Print.

Rapongan, Syaman (夏曼˙藍波安). *Memory of the Ocean Waves* (海浪的記憶). Taipei: Unitas, 2010. Print.

———. *Old Men of the Sea*. (老海人). Chunghe: INK, 2009. Print.

Soqluman, Nequo. *Tongku Saveq (Palisa)*. (東谷沙飛傳奇). Taipei: INK, 2007.

Tan, Amy. *The Joy Luck Club*. New York: Penguin, 1989. Print.

Vava, Husluman, Kadresengan, Auvini, & Badai. *Voices from the Mountain*. Trans. Shirley Shu-Hwa Wu. Nipomo, CA: Serenity International, 2014. Print.

Venuti, Lawrence. *The Translator's Invisibility: A History of Translation*. New York: Routledge, 1995. Print.

Yep, Lawrence. *Child of the Owl*. New York: HarperCollins, 1965. Print.

I

LANGUAGE POLITICS ACROSS ASIA

1

"Exophony" in the Midst of the Mother Tongue

Resources between Languages

Kazuhito Matsumoto

WHAT IS REPRESSED IN LANGUAGE LEARNING

What is that blankness I often experience when suddenly exposed to a foreign language or when abruptly required to use English, my second language? It is a question I have held in the back of my head since I gained the basic knowledge of English. Even a natural bilingual may possibly come across that blankness even for a split second at the moment of a sudden switch in his or her language. I suffer it also when I grope for some unlikely expression in my native language, trying to find the proper equivalence in translation for an English word or phrase. I used to deem, probably along with most language learners, that the sense of being lost in such blankness is a sign of linguistic incompetence—an ineptitude that should be overcome, eliminated, or at least ameliorated for smoother communication and more efficient translation. Which, however, I doubt now. I suspect that, once considered apart from linguistic efficiency and practicality, the phenomenon of being lost between languages should have its own significance and deserve deferential treatment rather than the dismissive one it seems to have received so far.

As if chiding the commonly held notion about expertise, Joseph Brodsky, an American émigré poet from Soviet Russia, underwrites the importance of linguistically being lost: "in the business of writing what one accumulates is not expertise but uncertainties. Which is but another name for craft. In this field, where expertise invites doom, the notions of adolescence and maturity get mixed up, and panic is the most frequent state of mind" (Brodsky 17). At least in literature, according to him, it is "uncertainties" rather than the flawless skill of expertise that should be relied on. He even asserts that "expertise invites doom," and that "panic" is the common state of writers. As if elaborating on Brodsky's view of writing, Yoko Tawada, a Japanese writer and poet based in Hamburg, explains in her collection of essays *Ekusophoni* (Exophony):

> If one settles down cozily in unquestioning belief in the naturalness of one's native language, no true interaction with that language can develop. Without breaking away from such belief, there would be no contemporary literature. I think, therefore, for literature, being outside one's native language is not some special situation. It is simply the moderate extreme of the usual situation. (*Ekusophoni* 114–15).[1]

"[B]eing outside one's native language" is losing one's linguistic bearings, and she says that it is the "moderate extreme" yet "usual situation" for contemporary literature. Uncertainties, panic, and losing one's sense of orientation—all these linguistically clumsy states of mind refer us to that blankness of being lost between languages, and as the two writers testify, they constitute the important part of contemporary literary activity.

I would like to use the term "exophony" to refer to this clumsy situation between languages. The term is usually used for the "phenomenon where a writer adopts a literary language other than his or her mother tongue" (Wright 2). Distinguished from those who are compelled to embrace a foreign language by either historical or political or circumstantial reasons, "exophonic" writers are those who choose to adopt it of their free will as the language of their writing. But I want to use "exophony" as a more general term, tracing it back to its etymological genesis, and make it mean the state of being between languages while moving "out of the familiar sounds of one's own tongue" (exo- + phony). "Exophony," therefore, is "not as simple as mastering one or another language, nor as facile as writing in more than one language" (Slaymaker 47).

Tawada has worked up this exophony of being in the realm between languages into her own literary strategy as poetics. She confesses: "In fact, I am not so much interested in the study of many languages itself. Rather than any particular language, it is the gap formed between two languages that seems important. I don't intend to become a writer of both language A and language B. I would rather find the poetic ravine separating two languages, and descend into its depth" (*Ekusophoni* 31–32). If my experience of that blankness at the moment of being suddenly exposed to a foreign language is an embryo of the experience of Tawada's "poetic ravine," such linguistically clumsy state of being lost in the realm between languages should have its own substantial significance to inquire into. In light of the exophonic realm thus to be exposed, the meaning of learning and teaching English as the lingua franca in Asia, specifically in Japan, should in turn be examined.

TRANSLATION AND EXOPHONY IN DOING ENGLISH IN JAPAN

"Doing English" has a long history in Japan as in many other countries in Asia. It goes back, for a conspicuous incident as its serious beginning, to the apparition of the four black warships in Edo (now Tokyo) Bay in 1853, led by Commodore Perry, threatening the shogunate government to end its isolation from the rest of the world extending over two centuries and open the country to trade with the United States. Japan was then woken up from its long sleep and began to devote itself to learn English to catch up with the West. English, however, was not the only language it chose for a means of its modernization. French and German were learnt eagerly as well. In the prewar higher education system, students were required to spend a considerable part of their energy and time to learn to read in some of these languages so that they should acquire the knowledge and wisdom of the West. Such foreign language education survived, though being gradually superseded by the demand for more practical language skills, even to the period long after English became the dominant foreign language in postwar Japan.

Significant results of such language education were not only the acquisition of knowledge and wisdom itself but also an enormous amount of works of translation accumulated in Japanese. Thanks to this, those who could read only Japanese were able to get an easy access to various fruits of the humanities and sciences of the West. But its more significant result was the formation of the modern Japanese language: Japanese should be reformed so that it could take over what were expressed in the Western languages. It was reshaped into a definite modern national language through such translations. Throughout this transformation, or more moderately, this adjustment of the Japanese language, English no doubt played the most distinguished role, especially in postwar Japan. In this sense, "Doing English" in Japan had much to do with "Doing Japanese."

Translation is assumed to be carried out between languages, each fully integrated and unified. Otherwise, how could one language be commensurate to another language in terms of transference of the contents from this side to that side? As the establishment of the modern Japanese language in face of English (and other Western languages) in the process of translation implies, however, a distinctive integrated language thus taken for granted is deceptive—it is the effect that translation produces rather than the condition it assumes (see Sakai 2). Just as the translations of the Bible into local languages helped establish European major national languages in the sixteenth and seventeenth centuries, the translation of texts in English and the other Western languages into Japanese served to reform it into a definite modern national language, which in turn instigated and created modern Japanese literature. Though there must have been a variety of frictions and conflicts as well as misunderstandings and indigestions involved in translation, the modern Japanese language thus established came to be considered by its users to be an independent authentic modern language comparable to English or any other Western language because of its ability and capacity to contain what those languages conveyed.

Minae Mizumura indicates the significance of the birth of a national language in modern Japan by referring to a symbolic episode from Natsume Sōseki's *Sanshirō*, a novel of the early period of modern Japanese literature. In *Sanshirō*, Professor Hirota chides a young man named Sanshirō, an imperial university student fresh from the country, who expresses doubt about his remark "Japan's headed for a fall," by pointing out that Tokyo is bigger than Sanshirō's home province Kumamoto, and Japan in turn is bigger than Tokyo, but what's bigger than Japan is the "inside of your head," therefore, he should not get bogged down by the nation. Mizumura elaborates on the episode:

> What makes the inside of one's head "bigger" than a nation? It is thinking in a national language that came into being through translation of *universal languages* [e.g., those then dominant European languages such as English, French, and German]. On his way from Kumamoto to Tokyo, Sanshirō encounters for the first time his own language used as a national language—a language that can transcend and critique one's own nation-state, a language that presumes a global consciousness. (Mizumura 141–42, emphasis original)

In light of the discussion so far, Mizumura's elaboration needs a further elaboration. Since a national language "came into being through translation," we should not ignore the scene of translating transactions. In translation there must be a moment when the "inside of one's head" is confined to neither of the two languages—when the translator's consciousness is in the state of uncertainty, being drawn to the outside of one's native tongue while powerfully affected by the foreign language. In this sense, the "inside of one's head," being conscious of the realm between the languages, is bigger than, or at least not contained in one's "national language." Such in-between realm of linguistic consciousness is the source for the renewal of one's own language as a "national language," and one should repeatedly experience being there lest the language thus reshaped be closed and become unable to "transcend one's own nation-state," being merely supportive with one's homogeneous language community.

Such renewal of the Japanese language through translation came to take a form of the interaction between Japanese and English, especially after World War II, resulting in stimulating and innovating contemporary Japanese literature. Its arguably most significant and recent example is novelist Haruki Murakami, an avid translator of American novels in his own right, always translating during intervals between his fiction writings. By the time he decided to take to a writing career, the modern Japanese language had established its own definite identity and become too burdensome for a would-be writer to find his own voice in it. He confesses that he wrote the first part of his debut fiction *Hear the Wind Sing* in English and then translated it into Japanese to obtain a "new style of Japanese":

> Since I was born and raised in Japan, the vocabulary and patterns of the Japanese language had filled the system that was me to bursting, like a barn crammed with livestock. When I sought to put my thoughts and feelings into words, those animals began to mill about, and the system crashed. Writing in a foreign language, with all the limitations that entailed, removed this obstacle. . . . Having discovered the curious effect of composing in a foreign language, thereby acquiring a creative rhythm distinctly my own, I returned my Olivetti to the closet and once more pulled out my sheaf of manuscript paper and my fountain pen. Then I sat down and "translated" the chapter or so that I had written in English into Japanese. Well, "transplanted" might be more accurate, since it wasn't a direct verbatim translation. In the process, inevitably, a new style of Japanese emerged. The style that would be mine. A style I myself had *discovered. Now I get it,* I

> thought. *This is how I should be doing it.* It was a moment of true clarity, when the scales fell from my eyes. (Murakami xiii–iv, emphasis original)

The "curious effect" he experienced while composing in English was caused not only by his distance from his mother tongue, writing in a foreign language, but also by his distance from that foreign language, not being its native user. He became a temporary exophonic writer, composing the part of his first novel in a language other than his mother tongue, thereby situating himself in the exophonic realm between the languages. The "curious effect" he discovered can be called by our term an "exophonic effect," which Murakami reinforced by translating what he had composed into Japanese. The exophonic effect thus experienced in the in-between realm of languages culminated in the emergence of a new voice of his own as a writer that could not be confined to the existing Japanese language.

What Murakami's superb case typically exemplifies is what English education in Japan had been secretly and rather hesitatingly valuing up until recently, in the teeth of the longstanding denunciation by the practical-minded of its translation-oriented pedagogy—not mastering English itself, not "doing English" *per se*, but experiencing the in-between state of Japanese and English. In view of this, it should be emphasized again, the seeming blankness or uncertainty one experiences often while trapped between languages—the rudimentary experience that may lead to Murakami's "curious effect"—must not be dismissed simply as a sign of linguistic ineptitude. We need to inquire into what is going on in the realm between languages, more in detail, in terms of exophonic experience.

THE IN-BETWEEN REALM OF LANGUAGES AS THE "SELF-FORMING FORMLESSNESS"

I would like to dive into what Tawada phrases the "poetic ravine separating two languages" by referring to an episode of her own. Tawada elsewhere tells of her crucial experience of being trapped between two languages when she began to live in Hamburg in her early twenties with only a little knowledge of the German language:

> Half a year had passed without speaking Japanese at all. It seemed to have left my life completely. I couldn't find a right name for anything I touched, nor could I come across any apt word to express what I felt. German didn't help me much, but no doubt about that, since it was after all a foreign language to me. To think of my mother tongue leaving me, I was scared as if its letters disappearing into a fog. Then I gradually began to feel, think, and make decisions without any help from language. Some would doubt it, but I actually managed without resort to language. Other than describing so, I have no way to explain the situation I was trapped into at that period. (*Katakoto* 11, translation mine)

What this radical exophonic situation added up to was the rebirth of both languages for her. As she recollects, "After a while, I started *translating* this wordless life of mine into either one of the two languages. It was as if, for instance, Japanese was reborn as a new body of language from its demise. My new born Japanese might be despised as sounding unnatural to some, but in any other Japanese, I felt like I was unable to verbalize what was going on around me" (12, emphasis original). Instigated by this experience of emerging from her wordless in-between blankness, she comes to doubt the "grounds for crediting so-called natural and beautiful Japanese while discrediting any other Japanese falling short of being qualified so," and concludes that "as far as those who live using it can verbalize their lives more genuinely, it doesn't matter if they speak or write in any such 'disqualified' Japanese." She even points out, "In this regard, perhaps, labeling it as 'natural or unnatural' and 'smooth or awkward' is just as discriminatory as political incorrectness" (12–13). The same goes for her German. Thus she began to write in German as well.

Tawada's extreme case discloses what lies beyond Murakami's balancing act between languages and reveals what he ultimately tapped into as the source of his "curious effect." She was deprived of the privilege Murakami enjoyed: she did not have adequate knowledge of the foreign language, nor was she backed up by circumstances where she could turn to her native language for the last resort. As a result, she got stuck in the "wordless life of mine" between languages, which she translated into a "new body of language." What was hidden behind Murakami's ingenious attempt of linguistic round-trip between the two languages is this wordless in-between blankness, from which he drew the "curious effect" for his "new style" of the Japanese language.

The blankness between languages as the source of linguistic renewal is also noted by Eva Hoffman, a Polish immigrant in Canada, in the course of her search for a "true voice," delving under layer of various acquired voices. In her memoir *Lost in Translation*, she reports that she discovers "silence" in the end:

> I reach what I'm searching for: silence. I hold still to steady myself in it. This is the white blank center, the level ground that was there before Babel was built, that is always there before the Babel of our multiple selves is constructed. From *this white plenitude*, a voice begins to emerge; it's an even voice, and it's capable of saying things straight, without exaggeration or triviality (Hoffman 275, emphasis added).

What Hoffman calls "silence" should mean, as with Tawada's "wordless life," a blank state of being devoid of any definite individual languages, but full of unfettered linguistic elements. Thus, she refers to it as "this white plenitude." The emergence of "a voice" from Hoffman's "white plenitude" must correspond to the emergence of "a new body of language" from Tawada's "wordless life" through what she calls translation.

Doug Slaymaker points out that "Tawada's characters [in her fictions] are constantly to be found in places where they are not comfortable, on the way to someplace else, rarely on the way 'home,' but in and out of languages, back and forth across borders" (Slaymaker 9). He goes on to note that they are dwelling on the "space between" that doesn't exist but in movement, "a space that is no place"— the "only place from which we can speak and write in the contemporary world" (9–10). Tawada herself has adopted the strategy of border crossing to put herself in such an exophonic state of being in the space between. Murakami also had once lived abroad, mainly in the United States, for years after his debut, to secure the "space between" for his articulating subject as a novelist in Japanese by crossing borders away from his native country and language. Our discussion on doing English in a non-English native country, or learning English in the midst of our mother tongue, however, compels us to be concerned, not with the geographical border crossing, but rather with the linguistic border crossing of going over the edge of one's own language into the uncomfortable realm of consciousness unconfined to any single body of language.

While mostly ignored by linguists and teaching a foreign language theorists, this uncomfortable realm of consciousness, or the state of being there as irreducible to any definite single language, has often been taken up by novelists, poets, literary critics, and philosophers as one of their main concerns. Walter Benjamin, for instance, in his classic essay on translation, as if referring to what Tawada and Hoffman call "wordless life" and "white plenitude" respectively, introduces the "primal elements of language" as some linguistic reservoir to mediate between languages in translation:

> The basic error of the translator is that he preserves the state in which his own language happens to be instead of allowing his language to be powerfully affected by the foreign tongue. Particularly when translating from a language very remote from his own, he must go back to *the primal elements of language* itself and penetrate to the point where work, image, and tone converge. He must expand and deepen his language by means of the foreign language. (Benjamin 262, emphasis added)

We must not overlook the point here that one's own language is expanded and deepened in translation, not directly "by means of the foreign language," but, first of all, by one's being "powerfully affected" by it and then "go[ing] back to the primal elements of language."

The exophonic experience of the realm unconfined to a definite language is not limited to those who are concerned about the realm between languages, but is also grasped by those who are simply anxious for plunging over the verge of their own respective languages. Kenneth Burke, an American critic and poet, for one, discussing linguistic transformation in one of his seminal books, delineates such experience by recourse to the image of a "great central moltenness":

> Distinctions, we might say, arise out of a great central moltenness, where all is merged. They have been thrown from a liquid center to the surface, where they have congealed. Let one of these crusted distinctions return to its source, and in this alchemic center it may be remade, again becoming molten liquid, and may enter into new combinations, whereat it may be again thrown forth as a new crust, a different distinction. (Burke xix)

Ralph Ellison, following Burke, inherits this "moltenness" in his classic black novel *Invisible Man* as the "chaos which lives within the pattern of your certainties." He makes the African-American protagonist go "underground" in order to overcome the problem of identity including the social "distinction" of black and white, where the protagonist-narrator taps the shared resources buried beneath social dichotomies.[2] After having stripped off his social-linguistic masks one by one, he decides to emerge from that chaos:

> In going underground, I whipped it all except the *mind*, the mind. And the mind that has conceived a plan of living must never lose sight of the chaos against which that pattern was conceived. That goes for societies as well as for individuals. Thus, having tried to give pattern to the chaos which lives within the pattern of your certainties, I must come out, I must emerge. (Ellison 580–81, emphasis original)

The narrator's plunging into the "chaos" and emerging from it with a new pattern corresponds to what Benjamin portrays as the translational experience through the intermediation of the "primal elements of language." Deleuze and Guattari, in their *A Thousand Plateaus*, sum up this dynamic relation between one's own tongue and the primal linguistic resources: "To be a foreigner, but in one's own language, not only when speaking a language other than one's own. To be bilingual, multilingual, but in one and the same language, without even a dialect or patois" (Deleuze & Guattari 109). In our terms, what is delineated here is called the state of "exophony in the midst of the mother tongue."

Deleuze and Guattari comprehend what is going on in this state of in-between as "becoming" (232), which they depict, leaning on Pasolini's *L'expérience hérétique*, as the "essential thing" to be found "neither in language A, nor in language B, but 'in language X, which is none other than language A in the actual process of becoming language B'" (117). John Krummel, in turn independently, drawing heavily on Japanese philosopher Nishida Kitarō, grasps this "becoming in language X" as the "alterity of the source of being and knowing that is always in excess to human conception," and describes it as the "self-forming formlessness as irreducible to form" (Krummel 46).

Finally, borrowing Krummel's term, I would like to conclude that what the articulative subject experiences through those linguistic renewals evinced by Tawada and Murakami as well as the others, undergoing exophonic experiences of being unconfined to any definite body of language, can be construed as the "self-forming formlessness" of linguistic consciousness. And the seeming blankness or uncertainty one often experiences when suddenly exposed to a foreign language is a gateway to this drama of the "self-forming formlessness" enacted by means of a foreign language between one's own tongue and the primal linguistic resources; therefore, it must not be dismissed simply as a sign of linguistic ineptitude.

ENGLISH AS THE UNIVERSAL LANGUAGE AND THE CONCEALMENT OF THE LINGUISTIC DRAMA

The drama between one's own tongue and the primal linguistic resources, readily accessible by way of a foreign language, is most likely to be dismissed as linguistic ineptitude in our time of globalism, when English is the indispensable equipment for individuals to compete and survive successfully in the world restructured by global capitalism, where the neoliberal individualism of self-discipline and self-responsibility is advocated.[3] One of the typical phenomena that resulted from such recent world shift is "audit culture." Audit is originally a term used in finance as accountant's inspection, but, as is often the case with neoliberalism, a business model of management has been imposed upon some other fields including an individual's practice, and it now means an assessment of individual performance as well as organizational one in light of accountability, transparency, efficiency, and effectiveness for self-governance.[4] Audit culture has long been imposed on Japan since the early 1990s when its bubble economy burst and it was forced to undergo drastic social as well as economic structural reforms: people, organizations, and institutions have been supposed, based on self-disciple and self-responsibility, to audit their performance in order to be competitive participants in the standardized world of global economy.

In such a society of audit culture, there is no room for one to be embarrassed between languages, for it is against the condition of being accountable and transparent required by its culture to wander off into some contingent situation without any definite goal in view. In audit culture, being powerfully affected by the foreign language one is trying to master and led to the ambiguous realm of the "self-forming formlessness" between the mother tongue and the primal linguistic resources is unthinkable, for audit culture, being concerned with the process for goal achievement, has nothing to do with something new, original, and unpredictable. English as the common language is an established fact in the world of global capitalism today; to be competitive in this world, therefore, it is an indispensable prerequisite to be a skillful user of English, comparable to its native speakers; being trapped between languages rather than being immersed in English, then, is a waste of precious time, an aberration from the efficiency and effectiveness needed to master the language.

Another ominous thing English as the universal language is threatening to cause is the loss of significance in writing for a national literature. Minae Mizumura sees a chilling omen for a national language and its literature in her provocative book *The Fall of Language in the Age of English*:

> We who have learned to read and write *in our own language* are no longer what we were. We have become deeply attached to reading and writing in our own language. Nonetheless, when one's own language circulates concurrently with a universal language in society, and when it is apparent that the latter is gaining force, an ever greater number of knowledge seekers will become bilinguals. In the process, how these bilinguals engage with their literature will inevitably be affected, which in turn will inevitably affect their literature—and their language. For the fall of language is first set in motion when these bilinguals unknowingly begin to read English and their own language on two different levels. (Mizumura 167, emphasis original)

"[O]n two different levels" she mentions is a foreseeable phenomenon that the bilinguals take more seriously what they read in English than what they read in their own language: for instance, "they turn to English-language media to learn about critical international events—they may or may not be conscious of the Anglophone bias there—and use the media of their own country only to find out the results of home sports games or follow home celebrity gossip" (168). And she shudders at the predictable future of her own literature: "Not only bilinguals but true readers of literature—not mere consumers of books—would eventually cease to expect their own language to bear the intellectual, ethical, and aesthetic burdens it once did" (168).

Mizumura does not deny the necessity of some intensive English education in Japan, but what she advocates is the necessity to "counter the mass hysteria that cries incessantly for more and more English" (198) and to reform the Japanese language arts curriculum for the protection of modern Japanese literature and its language. She asserts:

> But that all seekers of knowledge should use the identical language to think and to read and write is not a development to which humanity can remain indifferent. Reality is constructed by languages, and the existence of a variety of languages means the existence of a variety of realities, a variety of truth. Understanding the multifaceted nature of truth does not necessarily make people happy, but it makes them humble, and mature, and wise. (202)

I will accept this assertion of hers wholeheartedly, but I cannot dispel the suspicion that Mizumura is treating the modern Japanese language and its literature as something established and closed that should be cared and preserved rather than be affected, influenced, challenged, and changed. This suspicion seems to be confirmed by that, in her discussion on the fate of Japanese in face of English as the universal language, she never explores, or even mentions, the realm between languages, that realm unconfined to any definite body of language, that realm of "self-forming formlessness," of which the blankness or uncertainty one experiences often when suddenly trapped between languages is rudimentary. To pause on this clumsy condition of our linguistic experience would no doubt make us humbler.

EXOPHONY IN THE MIDST OF THE MOTHER TONGUE

One of the pleasures of learning a foreign language is to be led out of our own language into some unknown realm unconfined to either of the languages, whose virtue is that even an absolute beginner can experience it. I enjoy it whenever I have a rare chance to listen to poems recited in their unfamiliar original languages at some international poetry reading session. Semantically the poems usually sound almost completely meaningless to me, but once I shake off my obsession for meaning and yield myself to their acoustic materiality without restraint—that is, once I become "zero" with the poems recited by the authors, they begin to appeal to me through the affect and percept they yield for me. In the end, I am left with blocks of sensations, which yearn for expression in some definite form. This must be what Krummel terms as the "self-forming formlessness," or what Delueze and Guattari explain as the state of "becoming" to be found neither in language A, nor in language B, but in "language X," which none other than "language A in the actual process of becoming language B," or what they put simply as "to be a foreigner, but in one's own language."

As my precious experience of this kind testifies, there must be a goal in learning a foreign language other than that of mastering it. Once freed from the values and attitudes endorsed by global economy and its moral support of neoliberalism, we could possibly view the English language, not only as the language that should be mastered by everyone for the survival in the world today, but also simply as a foreign language that can be learned and taught to invite us into that exophonic experience of being lost in the realm between languages and being born again into a renewed body of language. Tawada describes the excitement of this drama of being lost and reborn as follows:

> It is wonderful to imagine that a new born baby is endowed with the latent capacity to speak all the languages on earth. But maintaining the capacity to speak them all means never being able to speak even one of them, which in turn means that, for a child to acquire a mother tongue, he or she should for the time being eliminate the possibility of speaking any other languages except the chosen one. . . . As adults we often feel inclined to learn foreign languages, which is probably because of our memories of the free and smooth movements of our lips and tongue as a baby. We talk and move our tongue a lot every day. But there must be movements of our tongue we no longer make and there must be places in our mouth our grownup tongue never touches. To read aloud haltingly a textbook written in a foreign language, feeling for those movements and places in our mouth, is an intriguing act of a tongue dancing, which could be an art. There dormant in me is a yearning for a tongue which is moving elastically in all directions, rolling back, clicking, and puffing,—a tongue, dancing around wildly for freedom, without forming any meaning at all. (*Ekusophoni* 47, translation mine)

The linguistic clumsiness of being trapped between two languages often makes it possible for us to be aware in ourselves of the yearning for a tongue that can dance wildly in the midst of linguistic resources, freed from the confines of the tongue we have already acquired. Exophonic experience might as well be defined as tongue dancing, exploring movements our tongue has never made and groping for spots in our mouth it has never touched, trying to catch unknown sensations and unheard sounds, and coming up with a new body of language. Pausing on the clumsiness of our tongue, being lost between two languages, is the gate that leads us to the linguistically unknown continent of our consciousness yet to be explored.

NOTES

1. All the citations in English from Tawada's works in Japanese are my translations, except for this and the next ones, which I rendered drawing upon the versions by Doug Slaymaker (Slaymaker 4–5, 5). I should also note here that *Yoko Tawada: Voices from Everywhere,* a col-

lection of studies on Tawada edited by Slaymaker, has confirmed my conviction of Tawada's relevance to the issue of "doing English in Asia."

2. For the detailed discussion of the relation of racial problem and the chaos, see Crable 168–74.

3. For the detailed discussion of global capitalism and its relation with global capitalism and neoliberalism, see Harvey 7–68.

4. For the overall discussion of audit culture, see Power, and Strathern.

REFERENCES

Benjamin, Walter. "The Task of the Translator." Translated by Harry Zohn. In *Walter Benjamin Selected Writings Volume 1: 1913–1926*, edited by Marcus Bullock and Michael W. Jennings. Cambridge: Harvard University Press, 1996. Print.

Brodsky, Joseph. *Less Than One: Selected Essays*. New York: Farrar Straus Giroux, 1986. Print.

Burke, Kenneth. *A Grammar of Motives*. Berkeley: University of California Press, 1969. Print.

Crable, Bryan. *Ralph Ellison and Kenneth Burke: At the Roots of the Racial Divide*. Charlottesville: University of Virginia Press, 2012. Print.

Deleuze, Gilles, and Felix Guattari. *A Thousand Plateaus: Capitalism and Schizophrenia*. Translated by Brian Massumi. 1987. New York: Continuum, 2004. Print.

Ellison, Ralph. *Invisible Man*. 1952. New York: Vintage, 1980. Print.

Harvey, David. *Space of Global Capitalism: Towards a Theory of Uneven Geographical Development*. London: Verso, 2006. Print.

Hoffman, Eva. *Lost in Translation: A Life in a New Language*. New York: Penguin, 1989. Print.

Krummel, John W. M. "*Basho*, World, and Dialectics: An Introduction to the Philosophy of Nishikda Kitaroō." In *Place and Dialectic: Two Essays by Nishida Kitarō*, translated by John W. M. Krummel and Shigeriori Nagatom. New York: Oxford University Press, 2012. Print.

Mizumura, Minae. *The Fall of Language in the Age of English*. Translated by Mari Yoshihara and Juliet Winters Carpenter. New York: Columbia University Press, 2015. Print.

Murakami, Haruki. *Wind/Pinball: Two Novels*. Translated by Ted Goosen. New York: Knopf, 2015. Print.

Power, Michael. *The Audit Society: Rituals of Verification*. Oxford: Oxford University Press, 1997. Print.

Sakai, Naoki. *Translation and Subjectivity: On "Japan" and Cultural Nationalism*. Minneapolis: University of Minnesota Press, 1997. Print.

Slaymaker, Doug, ed. *Yoko Tawada: Voices from Everywhere*. Lanham, MD: Lexington, 2007.

Strathern, Marilyn, ed. *Audit Cultures: Anthropological Studies in Accountability, Ethics, and the Academy*. London: Routledge, 2000. Print.

Tawada, Yoko. *Ekusophonii: Bogo no soto e deru tabi* [Exophony: Traveling outside the Mother Tongue]. Tokyo: Iwanami Shoten, 2003. Print.

———. *Katakoto no uwagoto* [Baby Talk in Delirium]. Tokyo: Seidosha, 1999. Print.

Wright, Chantal. "Introduction: Yoko Tawada's Exophonic Texts." In *Portrait of a Tongue*, by Yoko Tawada. Translated by Chantal Wright. Ottawa: University of Ottawa Press, 2013. Print.

2

Unfinished English

Stories from the Other Other

Will P. Ortiz

The Philippines was colonized by three countries namely—Spain, the United States of America, and Japan. Spain denied education to its colony and emphasized religion instead until the end of the nineteenth century; Japan briefly instilled Japanese values through education during World War II; and the United States of America educated the Filipinos in the beginning of the twentieth century. It is the American colonial education that has a lasting impact in the Philippines.

The first line of Renato Constantino's *The Miseducation of the Filipino* states that "Education is a vital weapon of a people striving for economic emancipation, political independence and cultural renaissance" (1). He asserts that, under the United States, no comprehensive educational program is given to Filipinos to advance political and economic emancipation. The past and present colonial and neo-colonial education of the Filipinos continue to adhere to the political and economic system of U.S. imperialism.

When the Americans conquered the Philippines in 1898, soon after a new system of education was imposed. Gen. Arthur MacArthur, Jr., an Army general serving as a Governor-General in the Philippines, recommended a large appropriation for education because he saw this as an "adjunct to procure and expedite the restoration of tranquility throughout the archipelago" (Constantino 3). Through the Department of Education under the Second Philippine Commission by the Organic Act of 1901, English was taught in the public school system. It was earlier instructed by U.S. President William McKinley in the First Philippine Commission that the local languages be used as medium of instruction but after two years, the policy was changed (Gonzalez 26).

English as a medium of instruction was seen as more effective in the pacification of the Philippine islands and because of this, more English teachers were needed. In 1901, the ship Thomas from San Francisco sailed to the Philippines with about five hundred American teachers. These first English teachers were popularly known as Thomasites. Gonzalez stated that there was a replacement pattern of teachers and by the end of Francis Burton Harrison's administration in 1921, "about 2,000 American teachers had come and gone" (26). Up to 1935, all the heads of the Department of Education were Americans.

The Hare-Hawes-Cutting Act and Tydings-McDuffie Law were passed allowing the independence of the Philippines in 1935 under a Commonwealth Government. An interesting provision of both was that English was to continue to be the medium of instruction in the educational system. One can only deduce that English as the medium of instruction in education had a positive effect (from the American point of view) in the "restoration of tranquility" as envisioned by Gen. MacArthur, Jr. Education was a tool in subjugating the mind of the Filipino, which effectively took control of cultural, political, and economic life in the Philippines. As Constantino noted, through American textbooks, Filipinos learned that the right education meant having the proficiency to speak and write in English. American knowledge, beliefs, and values were introduced and consumed to the detriment of Philippine culture.

Constantino further explained that a new alien culture brought by the new colonial masters was being adapted by Filipinos that would erase nationalism and enhance "miseducation":

> With American textbooks, Filipinos started learning not only a new language but also a new way of life, alien to their traditions and yet a caricature of their model. This was the beginning of their education. At the same time, it was the beginning of their miseducation, for they learned no longer as Filipinos but as colonials.
>
> They had to be disoriented from their nationalist goals because they had to become good colonials. The ideal colonial was the carbon copy of his conqueror, the conformist follower of the new dispensation. He had to forget his past and unlearn the nationalist virtues in order to live peacefully, if not comfortably, under the colonial order. The new Filipino generation learned of the lives of American heroes, sang American songs, and dreamt of snow and Santa Claus. (5)

He explained that new "economic attitudes" were fostered by American education. First, it presented an idealized picture of a rural Philippines that would strengthen the belief that the Philippines is only meant to be an agricultural country and not fit for industrialization. Second, the consumption habits were molded by the influx of American goods that led to the detriment of the Philippine economy. Education taught the Filipinos that everything good came from the United States, from its chocolates, films, books, textbooks, knowledge, and ideas. Philippine history was taught from the point of view of the Americans, emphasizing the role of American heroes while the Filipino revolutionaries were depicted as brigands. There was no violent annexation, but history taught Filipinos that there was a "benevolent assimilation" deemphasizing or even erasing the role of Filipino revolutionaries. Conquering the revolutionaries meant little dissent, which allowed the control of our economy and transplantation of American political institutions and ideas in the Philippines.

It is in the context of "miseducation" that the University of the Philippines (UP) was born. On June 18, 1908, by the authority of the United States, the Philippine Legislature passed Act No. 1870, which was also known as "University Act." It provided for the founding of the University of the Philippines. Section 2 of the Act states that "the purpose of said university shall be to provide advanced instruction in literature, philosophy, the sciences, and arts, and to give professional and technical training." Two years after, the Department of English and Comparative Literature (DECL) was born in 1910. DECL has since given birth to the College of Mass Communication, and the Departments of Filipino and Philippine Literature, Speech Communication and Theater Arts, and Arts Studies.

While the Americanized school system was being established, Pilipino nationalism continued. The Jones Law pledged the eventual recognition of Philippine independence in 1916, which created an all-Filipino legislature. However, the Jones Law still reserved the position of Chief Executive, and the head of Public Instruction (precursor to the Department of Education) to American officials appointed by the President of the United States. According to the Philippine government's Official Gazette, the latter was "considered the most important cabinet portfolio." The Tydings-McDuffie Act of 1934 provided for a ten-year transitional period to prepare for independence. These laws eventually led to the granting of independence. On July 4, 1946 Harry S. Truman granted independence to the Philippines.

According to Reynaldo Ileto (qtd by Gaerlan 191), "the critical decade for nationalism in the academy" was in 1955–1965. It was also during this time that the Rizal course was established. Republic Act No. 1425, mandated that the works and writings of Jose P. Rizal (the national hero of the Philippines), particularly his novels *Noli Me Tangere* and *El Filibusterismo*, be included in the school curricula whether private or public. The act prompted the translation, publication, and proliferation of the said novels in Tagalog and English. It was not clear in the law whether to teach it in English or Pilipino, but what was clear is that the teaching of Rizal underscores nationalism.

The Kabataang Makabayan (KM or Nationalist Youth) was founded by Jose Ma. Sison in 1964, which adhered to the Marxist-Leninist-Mao Tsetung political line. Sison was then a young faculty member in the Department of English. He would then become the head of the underground organization National Democratic Front. Writing in exile in 2013, he reiterates that KM was and is a revolutionary organization. "The KM mission was to continue the unfinished Philippine revolution against foreign and feudal domination and to support the working-class as the leading class in the new democratic revolution. The KM vision extended to a socialist future" (Sison). Sison was known in the UP as a subversive and attracted a large number of students in his organization. The period from the 1960s and the 1970s was a time of social unrest—demonstrations of workers, peasants and student activists proliferated.

The Department of Pilipino and Philippine Literature (DPPL) was born on January 27, 1966. "Pilipino" is used here to refer to the Philippine national language under the 1935 Constitution and "Philippine" is used to describe the different literatures from the regions and not just the literature written in Tagalog, the language used in Manila. Teachers from the DPPL came from the Department of English and Department of Linguistics. According to Gaerlan, the first chair Leopoldo Yabes was against the teaching of the Rizal course in Pilipino, emphasizing the use of English as a medium of instruction in the department.

In Teodoro's "The Department of Filipino and Philippine Literature: 1966–1981," aside from providing "advanced instruction in Filipino and Philippine literature and in the larger field of Philippine Studies" (2), one of the principal objectives of the DPPL is "to develop a deep sense of responsibility for professional growth through teaching, post-graduate studies, scholarly publications, and participation in national and *international conferences*" (3; author's emphasis). This meant that a faculty in the DPPL must not only be well-versed in Pilipino but also in English. Subjects such as Philippine Literature 30 (Philippine literary tradition), Philippine Literature 106 (Philippine Literary Criticism), would cover "native and foreign languages," "history, canons of and influences on the literature and criticisms in the Philippines" (3). It was in 1978 when two new courses were created—the Philippine-Southeast Asian Literary Relations and Philippine-Latin American Literary Relations.

In 1972–1973, Pilipino was introduced in the general education program. This was due to the growing nationalism in the university despite the declaration of Martial Law on September 21, 1972, under President Ferdinand Marcos. One of those arrested during this period was Ernesto Constantino in 1973, then the chair of the DPPL who advocated the use of Pilipino and not English, as a medium of instruction. Constantino's arrest was not due to his advocacy of Pilipino but because of the teaching of radical literature in the department (Gaerlan 201).

On February 28, 1974, the DPPL was changed to Department of *Filipino* and Philippine Literature (DFPL). "Filipino" is the new term for the national language based in Tagalog and inclusive of the contributions of languages in the Philippine regions, which according to McFarland, is about 110. "Pilipino" then was understood to be de facto Tagalog, which is the "dominant language not because of numerical superiority but by power, political, commercial or otherwise" (Llamzon, 1988, qtd by Shegina, 1998). Filipino is more inclusive and reflective of the nationalist fervor in the 1970s.

During the radical phase of nationalism in the UP, some of those known to be subversive members of the intellegentsia in the DFPL were Bienvenido Lumbera, Jose F. Lacaba and his brother Emmanuel Lacaba, and Renato Constantino. According to an interview with Pamela Constantino, a senior faculty from the DFPL, all of them taught at the Department of Filipino. All of them were writers in English that fought for nationalism. The first three were award-winning poets in English but would shift their writings in Filipino. Lumbera and Jose Lacaba were arrested and tortured during the Martial Law period. Renato Constantino's essay "The Miseducation of the Filipino," up to this day, remains to be a relevant reading in the DFPP curricula. The younger Lacaba, Emmanuel, was killed by the military in 1976. His works, compiled in a book called *Salvaged Prose*, is aptly titled because the word "salvaged" has a different meaning in the Philippine context. When a Filipino says, "Someone was salvaged," it doesn't mean someone was saved, it means someone was brutally murdered or was a victim of summary execution.

By recounting history, one can see how the curricula of the Department of Filipino was radicalized and how this affected the consciousness of students and the faculty. Always, literature was discussed according to its context. In the university, this author always heard the saying, "A text without a context is pretext." Literature is never neutral.

PARTISAN LITERATURE

Bienvenido Lumbera was conferred the honor of a National Artist for Literature in 2006. He is a literary scholar, librettist, poet, teacher and a political activist up to this day. His arrest during the Martial Law period never dampened his advocacy for a Philippines free from being a neo-colony of the USA. When this author asked for permission to reprint some of his poems, he answered in English, "Yes, but do not use the English ones." Activists during the Martial Law had deliberately chosen P/Filipino as the language to use in their writings because this was and is the language of the masses that would bridge the gap between the educated and the poor. The term "masses" does not just mean the "multitude" but the politically disenfranchised people.

His poems written in Filipino like "Lumen Christi (Light of Christ)," "Ka Bel (Comrade Bel)," and "Oyayi ng Yaya (Nanny's Lullaby)" can grip the consciousness of students because they can readily understand the persona. "Lumen Christi" uses the form of a prayer to illustrate the transformation of a privileged persona; "Ka Bel" salutes a friend/comrade whose life was dedicated to the masses he served; or the loneliness of an overseas contract worker who is forced to take care of someone else's child while she left her own baby behind in "Yaya's Lullaby."

Contrast these to the earlier poems of Lumbera like "To a Young Man Who Lost His Mind" with the persona of a colonial master talking to a man he colonized. Read the first four lines:

> My mind to me a kingdom is . . .
>
> Yours has fallen, and we can only smile away
> our shame when you invite armies less hostile than ours
> to spread their tents on territory no longer yours. (98)[1]

Another poem of Lumbera is "Pedagogy," which talks about pedagogy's futility. Read the last four lines of the poem "Pedagogy" and see how the image of the swirling water grasps the issue of futility:

the word is futile. We know
loss is rippled water.
whose geometry radiates
circles more vicious than death. (238)[2]

By reading his poetry in English, one can readily see that the assumed reader is not one from the masses who cannot understand the language. Lumbera's advocacy to bridge the gap between the masses and the intellectuals is the reason why he wanted to write and teach in Filipino. One can understand why he said, "Do not use the English ones." In a citation as a Ramon Magsaysay awardee, he was quoted as saying, "As long as we continue to use English . . . our scholars and academics will be dependent on other thinkers." One should "see Filipino literature through Filipino eyes." This is partisan literature.

Jose F. Lacaba, in his essay, "Why I Stopped Writing Poetry in English,"[3] states that he needed to address ordinary people in his poetry. "The washerwoman who did our laundry and the ambulant vendor from whom we bought sweetened soya snacks probably had no interest in poetry, but if it was possible to address them through a poem, I knew that the poem could not be in English." But before completely switching to writing in Filipino, he wrote his last poem in English titled "Prometheus Unbound."

Prometheus Unbound

I shall never exchange my fetters for slavish servility. 'Tis better to be chained to the rock than be bound to the service of Zeus.

—Aeschylus, *Prometheus Bound*

Mars shall glow tonight,
Artemis is out of sight.
Rust in the twilight sky
Colors a bloodshot eye,
Or shall I say that dust
Sunders the sleep of the just?

Hold fast to the gift of fire!
I am rage! I am wrath! I am ire!
The vulture sits on my rock,
Licks at the chains that mock
Emancipation's breath,
Reeks of death, death, death.

Death shall not unclench me.
I am earth, wind, and sea!
Kisses bestow on the brave
That defy the damp of the grave
And strike the chill hand of
Death with the flaming sword of love.

Orion stirs. The vulture
Retreats from the hard, pure
Thrust of the spark that burns,
Unbounds, departs, returns
To pluck out of death's fist
A god who dared to resist.[4]

This was originally published in *Focus Magazine* (July 14, 1973) under the pseudonym Ruben Cuevas. In his blog, Lacaba states that "Today, whatever standing I may have as a poet in the Philippines will probably be based on my Tagalog poems. But I will also probably be remembered, or remain notorious, for my last poem in English." Lacaba explains that the last poem is not entirely written in English. "It's an acrostic poem, and the first letters of the lines, if read downward, spell out a Tagalog slogan popular among demonstrators before martial law: MARCOS HITLER DIKTADOR TUTA (Marcos Hitler, Dictator, Running Dog)." Up to this writing, "Prometheus Unbound" can be read in his blog. This is partisan poetry.

What then is said of poetry with no partisanship?

One of the most anthologized Filipino writers in English is Jose Garcia Villa. E. San Juan explained that when Villa arrived in the United States he was already called a "modernist master" (452) and with the publication of his books *Have Come, Am Here* (1942) and *Volume Two* (1949), his poetry earned praises from the leading critics such as David Daiches, Dame Edith Sitwell, Elliot Paul, Marianne Moore, and Babette Deutsch. He has been a recipient of a Guggenheim fellowship, an American Academy of Arts and Letters award, and a Shelly Memorial Award for Poetry. In the critical essays included in *Poems 55* by Villa, Daiches compared him to Blake and Yeats, "Villa is unable to make sense out of what he sees or feels until he has constructed a myth out of it (44)." Sitwell called one of his poems "a strange poem of an ineffable beauty" (45). Paul called him a poet with an "idiom so subtle and so personal he can get away with murder" (50).

"Lyrics 15," sometimes called "First, a Poem Must be Magical," is probably the most recognized poem by Jose Garcia Villa in the Philippines. The poem makes the reader hear the musicality and see the beauty of poetry even in just eight lines:

Lyrics 15

First, a poem must be magical,
Then musical as a seagull.
It must be a brightness moving
And hold secret a bird's flowering
It must be slender as a bell,
And it must hold fire as well.
It must have the wisdom of bows
And it must kneel like a rose. (12)[5]

No one can probably deny the greatness of the craft of Jose Garcia Villa while studying the form. But by considering the context of the poem, the usual discussion turns to the uselessness of aesthetics in the face of Philippine poverty. "What is the use of beauty when the masses are starving?" is a question that needs to be asked in the classroom setting. In the scale of partisan poetry, the poem loses its weight.

E. San Juan looks at Villa in a different light, seeing the critics' endorsement of Villa as the annulment of the Filipino race by erasing the poet's subject position. For example, E. San Juan notes critic S. E. Solberg saying that Villa's poetry is "personal and idiosyncratic fable, a protean version of the 'making of Americans'" (453). To San Juan, Villa's poetry can be read as the "subjective expression of a social antagonism" (453) but American critics have erased such readings because Villa's poetry is only seen as "personal." Such a reading is so different from the usual orientation of reading Villa as non-partisan.

Another poet who is often anthologized is the late Alfrredo Navarro Salanga. Salanga was a poet, journalist, novelist, playwright and fictionist. According to his daughter, Elyrah Salanga Torralba, he was also a member of International PEN, recipient of the Mulry Award for Literary Excellence and Ten Most Outstanding Men of the Year (TOYM) for Literature and Journalism and other awards. Salanga was also one of the founders of Philippine Board of Books for Young People, Manila Critics Circle, and Philippine Literary Arts Council.

Salanga's works that were socially committed include *Writings in Protest 1972–1985*, (1993) and *The Birthing of Hannibal Valdez* (1984). Torralba explains that he was detained twice because of his writings against the dictatorship during the Martial Law, once in December 1972 and again in November 1976. His poetry captures the anti-colonial sentiment of nationalist Filipinos. One such poem has Edwin Thumboo's name, a Singaporean who's known for his poetry with nationalistic themes. Like Salanga, Thumboo wrote in English. Read the entire poem and see how the poet gives voice to the suffering, hungry people who see poetry as silk:

For Edwin Thumboo and All of Us Who Suffer Through English in Asia

A travesty, they say,
a tapestry, we contend,
as we worm verses

        into languaged
silk.

        Or silt,
they say, these coccoons
are empty anyway—

        Cotton, they add,
        cotton
is the cloth of the people.

Silk is as good as silt to
them,

    butterflies a luxury
    because inedible,

except in extreme cases
because hunger

    gives them
    strange appetites.

Poetry is as silk is
just as novels are, for that matter,
as pretty flowers
on a table.

When empty, the flowers fade
into other meanings:
"Poet,
can verses in your language
feed me?"

Can your verses
    cocoon
my fear? Can it
shut off
    the howling
    of my children,

their hungry mouths
wide as mothwings,

the wick
of my lamp gone dry?

"Poet,
can verses in my language
feed me?"

Can they build
    a cocoon
large enough
    to weave
my anger?

    Or
will your verses

remain
    food for the few
who love tapestries
who love silk
who love butterflies

and flowers
on their tables?[6]

In Salanga's "A Philippine History Lesson," it depicts a Philippines with a distorted history brought about by its first colonizer, Spain. This distorted history is further deteriorated when the second colonizer arrived and miseducated the Filipino who considered American colonization a benevolent one. The poem attests to a chewed and dried history:

A Philippine History Lesson

It's history that
moves us away
from what we are

We call it names
assign it origins
and blame the might

That made Spain right
and America—bite.

This is what it amounts to:
we've bitten off, excised
from the rind of things.

What once gave us pulp
has been chewed off
and pitted—dry.[7]

What is emphasized in most of the readings in UP is the sense of history and how it was "pitted dry" and distorted by our colonial masters. It underscores the role of the privileged educated few in bringing socio-political changes together with the disenfranchised masses.

The history of colonialism, the subsequent suppression of nationalist movements, the declaration of Martial law and the state's perpetuation of human rights violations, the continuing social unrest, the neo-colonial policies that keep most Filipinos below poverty level—all contributed to the nationalist orientation of the students and faculty of the University of the Philippines. It radicalized the curriculum of the studentry. It emphasized a nationalist education that will serve the poor, an education that goes beyond the four walls of the classroom to see what is life outside the university. Thus, *Pedagogy of the Oppressed* by Paulo Freire gives importance to an education that critiques and questions the powers-that-be and the banking system of education; *Theater of the Oppressed* by Augusto Boal transforms the theater as an arena for liberation. The book, for the author, speaks volumes on why Shakespearan plays alienate a viewer from the Philippines and why it is necessary for plays to be translated; *Sociological Imagination* by C. Wright Mills opens the eyes of the reader about how a person's private trouble is connected to his/her social milieu, and how structural changes affect the private life; Althusser's "Ideology and Ideological State Apparatuses" emphasizes the tacit ways of how people imbibe the ideology of different institutions. In discussing hegemony, "gahum," a native Filipino term is used: literatures from Southeast Asia such as Indonesia, Malaysia, Vietnam, and Thailand are read, especially those that were written by authors during the colonial and post-colonial period. Writers who resisted colonization like Chairil Anwar, David Diop, Sahir Ludhianvi, Pablo Neruda, and Cesar Vallejo are all read and translated into Filipino. These poets spoke to us as a culture resisting colonialism.

Latin American writings such as Gabriel Garcia Marquez's *One Hundred Years of Solitude*, and Isabel Allende's *The House of the Spirits* are must-reads before a student graduates. These magical realists are read to supplement anti-imperialists short stories in Filipino. Magical Realism is a convenient term Filipinos use, different from how the Western world defines it. While the West sees magical realism as a mixture of fantasy and reality, to us, there is no magic to be distinguished from reality in the novels of Allende or Marquez. If the novels are peppered with the ghosts of loved ones, for us, it is not something that defies the natural/physical laws but something that occurs in our everyday lives. As a culture, we believe in ghosts, we believe in exorcising dark elementals that lurk in the dark corners of universities, we believe in *babaylans* or shamans, we believe in evil spirits, and we also believe in condemning evil (e.g., disappearances and torture) perpetuated by the neo-colonial state. Magical realist novels capture what we experience as a nation that experienced colonization. Perhaps, by seeing the supernatural as natural phenomena, we as a culture defies the "rationality" brought by colonization and neo-colonization.

If there are writers who are canonized in the academe, there are also Filipinos writing in America who are considered to be un-Filipino. This is the kind of writing that mimics being Filipino but stands out as a writing that sees Filipino as the other. Jessica Hagedorn's *Dogeaters* is one such example of Filipino othering despite the superfluous praises received by the author.

OTHER'S OTHERING

"This is a novel pretending to be Filipino. Tell me why," said my Filipino teacher while brandishing a copy of Jessica Hagedorn's *Dogeaters* back when I was an undergraduate student. Needless to say, Hagedorn's novel was not received well in class. If the novel has any political meaning, it has been drowned by Hagedorn's postmodern stylization. E. San Juan says this about Hagedorn's novel:

> In trying to organize and make sense of the fragmentation of the comprador-patriarchal order which sacrifices everything to acquisitive lust, she resorts to pastiche, alienatory montage of diverse styles, cliches, ersatz rituals, hyper-real hallucinations—a parodic *bricolage* of Western high postmodernism—whose cumulative force blunts whatever satire or criticism is embedded. . . .(455)

Carolyn Hau traces the literary mode of production of Hagedorn's novel. Hau points out the fact that *Dogeaters* was written by a "Filipino American through the grants and endowments of American institutions, published in the States, read by Americans and reviewed by American critics and writers (116)." She reads Hagedorn's project as the "Othering of the Other of the West" by virtue of Hagedorn's position as a privileged Filipino-American living in America looking into the affairs of the Philippines and seeing the dogeaters, "alleged to be a slang term for Filipinos" (122) as the other.

Rio, one of the narrators in *Dogeaters*, has autobiographical similarities to the author, Hagedorn. She attempts to understand her past by going back to the Philippines but she goes back to the United States, her real home, still disconnected to her nation of the past, a nation she never really knew. Rio failed to reunite with her past because she chose to do so, she chose to see the evils of her country—the decadence of a dictator and the First Lady, corrupt politicians, human rights violations, poverty, middle-class immorality and political turmoil. These are just broken fragments of her imagination of a nation she once knew.

While Hagedorn attempts to reconstruct her past, her identity and nation, she creates a novel that pokes fun at the Philippines at the expense of Filipinos with no dual citizenships, no means to travel from one country to another with the freedom money brings, no other country to escape to (like her characters) when there is coup d' état, martial law and political unrest. Hagedorn uses McKinley's writings, Jean Mallat's historical document in the nineteenth century to catch snippets of Philippine history from the point of view of colonial masters. She depicts characters with no redeeming value from Joey, a half-African-American, half-Filipino gay prostitute who witnesses the assassination of Senator Avila, the First Lady who is very much fascinated with herself and enthralling foreigners to her beautiful Philippines to the detriment and death of poor laborers, to Freddie Gonzaga, Rio's Father, who believes in dual citizenships and dual passports, to Rio's mother, Dolores Logan Gonzaga, who has an American father, is ashamed of her Filipino blood and is preoccupied with the preservation of her youthful beauty, to even Rio, her daughter, who is an unreliable narrator unable to recall her past clearly.

In the last chapter titled "Kundiman" (Tagalog love song), Hagedorn recreates not a love poem or song but a blasphemous mishmash of prayers without intending to provide an ending to the major characters' dilemmas. Hagedorn left the characters hanging. The author is like a tourist, but visiting only the ugly, dark, violent scraps and fragments of the Philippines and went home to write about what she saw. She takes some pictures of Philippine reality, (such as the tortures committed by the Martial law regime, the religious fanaticism of Raul Gonzaga, of a gay prostitute, of a beauty queen becoming a rebel) and then creates her own Philippine realities from the position of an outsider masquerading as an insider with an explosive grand truth to divulge to the western world.

As Filipinos living here in the Philippines, we can't understand how American critics can see the book as funny. By Rio distinguishing herself from the struggle and political turmoil in her birth country, and choosing to go home to the United States, not wanting to be a part of the change or transformation, she no longer can call herself a Filipino. Much like what the author Hagedorn did. The title *Dogeaters,* a derogatory term used by Americans became a word used by Hagedorn to describe a race that she no longer has any connection to, thereby othering the Filipinos living here.

Contrast this to the novel *Noli Me Tangere,* a required reading for all students in the Philippines, which also has the theme of returning home, identity and nation. Originally written in Spanish in 1887, it has since been translated into many languages. The recent translation in English was done by American writer Harold Augenbraum, published by Penguin Classics in 2006. In this novel, the wealthy hero Ibarra (born in the Philippines with half-Spanish, half-Filipino blood) returns home after acquiring his education in Europe. Seeing that his fellow Filipinos lived in poverty, superstition, and blind faith in Spanish friars, he wanted to put up a school, believing that education will pave the way to his country's progress. But since the Philippines was under the power of friars, not the governor-general, his efforts were in vain. At the end of the novel, due to political intrigues concocted by the friars, he was hunted by the Spanish authorities. A revolutionary named Elias who sees Ibarra as the hope for the transformation of his country took the bullet for him. Ibarra was able to escape. In the sequel, Ibarra will return to the Philippines with a different identity.

While this novel does not end well for the hero's goal, the concept of returning home to bring about changes in one's nation is clearly seen. The hero is in the center of conflict and does not choose to be a spectator and then escape when there is great political and economic unrest. The novel demonstrates the core values of *kapwa* and *damay* in Filipino; roughly translated, the values refer to having a sense of solidarity or community with one's fellow Filipino. Hagedorn's novel, sadly, is far from demonstrating such values.

While a lot of Filipinos do not speak and write in English fluently, which creates a cultural divide between the masses and the educated, many recognize the fact that there is a need to publish and write in English too. Writing in English does not necessarily alienate the Filipino. It is the content that matters. Nationalism does not speak in one language alone and there is a need to address the reading public in the national language and English.

CHANGING TIMES AND BILINGUALISM

As part of the changing times, publishers are now opting to publish picture books and poetry in Filipino and English. Rio Alma, aka, Virgilio Almario, a National Artist for Literature, published his book in 2008, *Isang Sariling Panahon: mga Tanaga, Diyona, Dalit, at Tatlong Talingdaw (A Time of One's Own: Poems in the Traditional Short Filipino Forms of Tanaga, Diyona, Dalit, and Three Talingdaw)* using precolonial forms of Tagalog poetry.

Rogelio G. Mangahas' *Gagamba sa Uhay* (Spider on Stalk) made use of the Japanese haiku in his Tagalog haiku and published it with translations. According to Virgilio Almario in the introduction of the book, the poet aptly titled the book "Gagamba sa Uhay" (Spider on Stalk), poem 115 in the collection, because this haiku is the "ubod" (xxvi) or heart of the collection. The grand theme of life and death is highlighted in the poem. The 270 haikus showcase Mangahas's prowess as a great Filipino poet.

In the field of children's book publishing, Adarna Books, Lampara Books, Tahanan Books, Anvil Publishing are just some of the publishers that chose to publish in two languages, (i.e., Filipino and English). While browsing the picture books of different publishers, this author noticed that texts written in English and Filipino do not have the same font size. Perhaps, it depends upon the original material the author submitted. If the original material is in Filipino, the text in Filipino is bigger and vice-versa. Why publish bilingual children's books? In this time of globalization, most middle-class parents who buy these books want their children to learn English at an early age.

But what kind of English do Filipinos speak? Most middle-class Filipinos would not recognize the different Englishes spoken here and abroad, but some would question the way Filipinos speak English, implying that there is only one correct way to speak English, that is, American English. The debates usually arise in the academe and Filipino writers in English. Philippine or Filipino English shall be tackled next.

PHILIPPINE ENGLISH/ FILIPINO ENGLISH

In a writer's forum tackling Philippine English, Marjorie Evasco, a poet writing in English and Cebuano, was asked by a British fiction writer to change one of the phrases of her poem "Acquainted with Lightning," because the phrase "raw blood" sounded "redundant or superfluous to the image" (Bautista 164), therefore "raw" should be deleted. The British writer was shocked with Evasco's explanation ". . . in my culture, animal blood can be cooked and eaten with the innards in that Filipino delicacy we call *dinuguan* or in my region, *dugu-dugo*. The adjective raw stayed and the poem has since seen publication in the *New Strait Times* in Kuala Lumpur and the literary journal in Southeast Asia called *Tengara*" (164).

F. Sionil Jose, a novelist known for his books like *Po-on* (1984) *The Pretenders* (1962) *My Brother, My Executioner* (1973) and *Mass* (1974), says that he writes primarily for Filipinos. In the same forum, he explained why he never uses the word "summer." "I have expunged the word 'summer' from my writing unless it is in the context of the four seasons. . . . Your readers elsewhere will not only misunderstand you but more than this, you are not true to the language itself. Because there is no summer in this country. We have a dry season, wet season, rainy season, dusty season but never, never 'summer'" (Bautista 168).

Jose further explains that he uses what Filipinos know firsthand without further need of explanation. "Take for example the balete tree. It's been there as a very good metaphor, but nobody ever used it. Because most of us are attuned to Greek mythology, we are not attuned to things right before us. I don't say 'apple green' anymore, I say 'mango green.' I don't say that the object was 'brown,' I say 'chico brown,' because I'm writing primarily for Filipinos" (168).

He concluded that international writing also means adhering to the standards of the Western world. To him originality meant excellence in writing.

> Those writers who feel that we should write internationally, well and good. But much of our writing to my mind is not only derivative but also not very, very well done. It is only when we can be original, when we can also write with excellence, that we can hope for readers not only here but also elsewhere in the English-speaking world. (169)

At the conclusion of the forum, most of the writers agreed that they use whatever words are necessary to convey meaning and do not worry about English standards of international editors. If most of the writers do not conform to international standards, is there a Filipino standard for English?

What then is Philippine English or Filipino English? Both of these terms are used interchangeably here. Llamzon defines what he calls standard Filipino English. "*It is the type of English which educated Filipinos speak, and which is acceptable in educated Filipino circles.* It is not the creolized form of speech spoken in some areas of the Philippines; nor is it erroneous attempts to speak in English . . . when Filipinos finally gain proficiency in English, they speak a homogenous type of English" (15, emphasis original).

Filipinos use brand names to cover the whole category. The most common terms are frigidaire for refrigerator, colgate for toothpaste, scotch tape for adhesive tape, tupperware for plastic products used for storage, milo for chocolate drink, zonrox for bleaching products used for cleaning, band-aid for adhesive bandages, skyflakes for crackers, pampers for diapers, kodak for camera or kodakan to mean taking pictures, crayola for crayons, modess for sanitary napkin, cream silk for hair conditioner, osterizer for blender, cutex for nail polish, citizen for wristwatch, johnson's for powder used for babies and sometimes, even coke for soft drinks.

Bautista, in "The Lexicon of Philippine English" cited nouns that many educated Filipinos turn into verbs. These are often used by the media, whether in print or broadcast media. "The verb *fiscalize*, to call attention to abuse, comes from the noun *fiscal* and does not appear in either the Macquarie or Webster's. . . . [T]he word *conscienticize* is used in the Philippines and this has probably come from the theology of liberation of Latin America" (54).

Neologism or inventing a word or a phrase is also a part of Philippine English:

> On the analogy of employee, examinee, grantee, Philippine English has *awardee, honoree, mentee* . . . Philippine English uses *jubilarian. . . . Reelectionist* and *rallyist* do not appear in other English dictionaries, nor do the words *bedspacer, carnapper, cockfighter,* and *holdupper. . . .* [T]he first postgraduate degree may be called a *masteral degree* on the analogy of "doctoral degree." *Studentry* is used in Philippine English. (59)

Often, visiting foreign friends find the following words so funny. "Presidentiables" (candidates for presidency) and "vice-presidentiables" (candidates for vice-presidency) are now consistently being used in the media. "High blood" is an adjective for someone who is angry or even mad. "Bath" means a shower or "bathing" using the traditional pail and a plastic scoop or dipper. "Carabao" (from the Tagalog "kalabaw") is often used for water buffalo, and "viand" for any kind of dish Filipinos eat with rice and not necessarily delicious nor tasty. "Motel" is a cheap, small hotel (not necessarily a motor hotel and not necessarily clean) used for premarital or extramarital sex. Similarly, the word betamax was used to mean pornographic movies and not just the videocassette tape. Today, the word betamax refers to a grilled congealed pork or chicken blood shaped like a small betamax sold in the streets. Other street foods named after brands are adidas for grilled chicken feet, and walkman for barbecued pig ears. The word "trapo" is used to describe *tra*ditional *po*liticians or politicians who are corrupt or dirty. It also has a double meaning because in Tagalog, "trapo" means a filthy, disposable rag. These words are used not just because of inventiveness but because as a people, Filipinos find humor in everyday life despite the poverty and other pressing political problems.

Filipinos use a lot of acronyms for a lot of phrases. CR or comfort room means a toilet or bathroom. OA means overacting. OP or out of place describes not a thing but a person who sticks out like a sore thumb or maybe a third wheel. MU or mutual understanding is the relationship between two people romantically involved but with no obligations or commitments. TH or trying hard describes a person who is trying too hard but failing at the same time and is tantamount to being a loser. DH or domestic helper became popular because of the burgeoning number of Filipino migrant workers who are usually women and most of the time it is used derogatorily. The government once used OCW for Overseas Contract Workers, but today the government popularized OFW to refer to the Overseas Contract Workers, which includes the DH.

The reduplication pattern in Filipino languages can be clearly seen in the way Filipinos speak English. Back in 2010, a Filipina beauty titlist competed in the Ms. Universe pageant. She was asked if there were any big mistake that she has done in her life and how she was able to correct it, and she said "There is no major, major problem that I have done in my life." While the American media criticized the content or the lack of content of her answer, Filipino netizens cringed with her grammar. The "major, major" answer was highly publicized here in the Philippines and became a standing joke. While most Filipinos would criticize her grammar, this writer would think that this is an instance of Filipino English. Filipinos usually say, "I feel very, very glad" instead of just saying "I feel very glad." Filipinos have to put the second "very" for emphasis. Instead of saying "I'll see you Tuesday after next," Filipinos would say "I'll see you next, next Tuesday." In the same manner, "last, last Christmas" means "the Christmas before last."

This reduplication in Filipino language can also be seen in the way we borrow English words. "Haping-hapi" means "very happy" with the "y" respelled as "i." The *ng* is the linker. Filipinos say "bising-bisi" to mean "very busy." "Type na type" means liking something very much whether it is a thing or a person. The *na* is the linker. "Miss na miss" means missing someone or something very much. Filipinos always use "love na love" to mean loving something or someone very much. Of course, these words are not used in the academe but in everyday life.

Filipinos have a different syntax. In the Philippines, the word "already" is used at the end of sentences. We say "I'm finished already," "I read it already," or "I'm eating already" instead of "I'm already finished," "I already read it," or "I'm already eating." This is because in Tagalog the word "na" is always used (which when translated in English becomes "already") at the end of a sentence to emphasize the time when something is being done or already done. Sometimes, instead of using "already," we substitute the Tagalog "na." It is equally accepted to say, "I'm finished *na,*" or "I read it *na,*" or I'm eating *na.*"

The Tagalogs will also use "ba" at the end of interrogative sentences similar to "na." Instead of saying "Is it raining?," it is customary to say, "Is it raining *ba*?" The *ba* is a Tagalog term used as tag question/marker. It is used for emphasis. "Did you miss me *ba?"* may mean "Did you really miss me?" or "Did you study *ba*?" means "Did you really study?" Tagalogs can also say "Did you study *na ba*?" as a substitute for "Did you already study?" or "Have you eaten *na ba*?" for "Have you already eaten?" It is redundant and certainly not needed in English sentences but a Tagalog feels that something is missing without adding the *ba* or sometimes the *na ba.*

Tagalog honorific words or phrases are also added when using English to convey respect. "Yes *po,*" shows respect when talking to someone older or someone with a higher status than the speaker. "Po" is a polite yes. It is very hard for a Filipino to just say something in English without using any honorific term. The Tagalogs will say, "When will you deliver *po*?" to be polite. Without the "po," the sentence sounds like a command (and the delivery might get delayed). In other non-Tagalog regions, they use honorific terms when talking in English such as "Manang," (a term used for a woman older than you) or "Apo" (used when talking to a grandfather or an elderly man). Instead of just saying, "When are you leaving?" one says, "Manang or Apo, when are you leaving?" Not to do so would be disrespectful.

Often Filipinos use "I'll go ahead" to mean "Bye, I'm leaving before you," and it does not mean that one is talking to someone she or he is rendezvousing with later. This will certainly create confusion when Filipinos talk to foreigners. "For a while" is used when answering the phone and it means the caller is being asked to wait a while. This author has never heard of anyone using it other than in a telephone conversation. Saying "Where are you going?" is a friendly way to greet someone. Foreign friends might think Filipinos are being nosy by asking this but they really don't want to know where you're going. This is just like talking about the weather. The answer to "Where are you going?" is "Diyan lang" or "Just there," and that answer satisfies Filipinos who understand that you really don't want anyone to know what you're doing and where you're going.

Many might disagree with the above examples as part of Filipino English, but there is no standard Filipino English just yet. And this writer doubts if this would happen in the near future. How can one standardize the way Filipinos speak English? Not even the academe can do that. The power lies within the majority of Philippine English speakers. But it is interesting to note that while some Filipinos accept Filipino English, some do not. In Ong's *Equipments, Pictorials and For a While, 284 Common Mistakes in English Made by Pinoys,* Ong does not accept the use of brand names as generic terms and other common Philippine English terms.

In class, one of the students of this author said that she read *The Handmaid's Tale* by Margaret Atwood and there she read the word "salvaging" used by Atwood with the same meaning as the way Filipinos use it. Upon checking Atwood's dystopian novel, "Salvaging" appeared in part XIV and XV and its meaning is summary execution. Atwood was aware that she borrowed the term from the Philippines. In the introduction by Bloom in his study of Atwood's work he stated that according to critic Coral Ann Howell, before writing *The Handmaid's Tale*, Atwood read a lot of materials, which include files coming from different countries including the Philippines (13–14). In the last part of the novel, Atwood's character Professor James Darcy Pieixoto explains that the term originated from the Philippines. Atwood, a Canadian writer, used Philippine English.

Whatever the language editors or language purists say, it is the people who use the language that will determine whether Philippine English is acceptable or not. At this point, it seems that Philippine English is gaining ground. Refusing, consciously or not, to speak English the way the colonial masters speak English is one way of resisting colonization. It is a way of owning what one has learned without being a caricature of the colonizers. Owning a language means having the power to think, write, and speak the language without resorting to mimicry. Owning the English language means having the right to determine one's own culture and, eventually, the right to determine one's own future.

If there are disagreements with the use of different Englishes, there are also contending opinions in the use of Filipino and English in the academe, which has escalated into more heated debates. The Philippine government sees it fit that the medium of instruction in education should be in English, relegating all other languages in the Philippines as second class. In the changes in the curriculum, there is a token appreciation of Philippine languages but the emphasis of it is still in the use of English for communication. The politics of language will be discussed next in the context of globalization.

UNFINISHED ENGLISH

As a teacher in the Department of Filipino, I do not deny the need to read literary canons in English. Teaching Philippine literature and creative writing in Filipino meant reading the canon of Anglo-American Literature and literary theories; reading the Latin American, Southeast Asian and African writers; reading Filipino works in English; and reading Filipino writers in Filipino. In fact, while browsing through the must-read texts for students, approximately, more than half of what the students read is in English. They haven't read enough in Filipino. This may be the reason why students have a hard time reading in Filipino; many of them see it as having so many "deep words" in Filipino or hard-to-understand words.

Issues surrounding these are the translations in Filipino that are even harder to read than the original English text. One example would be a student reading *God of Small Things* by Arundathi Roy rather than reading the translated Filipino version, *Diyos ng Maliliit na Bagay*, translated by Monico Atienza. Students are not equipped to read in Filipino because they come from middle-class families who read and speak English at home. Most middle-class parents speak to their children in English because it is the language needed to be gainfully employed in the future. Like most middle-class Filipinos, they were Disneyfied and McDonaldized consumers of an alien culture made familiar by global Americanization. Filipino middle-class families often buy books mostly written by Americans and published by American publishers and consume DVDs and movies made by Americans. Not all parents would see the need to read Philippine literature written in Filipino nor watch movies made by Filipinos. In college, while cultural imperialism is often discussed through Philippine cinema and pop culture, the analysis *mostly* boils down to its foreign sources, influences, and theories. But this does not mean that criticism and theorizing can only be done by imbibing foreign theories.

Gonzalez in his article, "The History of English in the Philippines" believes that the intellectualization of the Filipino language can only rely on American scholarship:

> Whatever one's feelings about it, one foresees that English will be the source for the intellectualization of Filipino, for as Filipino develops as a language of academic discourse, Filipino academics will have to continue to rely on American sources for their scholarship and act as translators and popularizers of knowledge in the local language. (Gonzalez 35)

American sources are *not* the only sources of knowledge. The fact is that the Department of Filipino keep on reading text and theories using the Filipino point of view. There are debates, but still, the Filipino theories of Bienvenido Lumbera, Virgilio Almario, Virgilio Enriquez, Prospero Covar, Zeus Salazar are just some of the essential readings for students enrolled in Philippine studies. These Filipino authors are required to be read by students in the University of the Philippines. These are necessary to combat the onslaught of globalization emphasizing the need to use English to the detriment of using Filipino as an intellectual language.

Globalization's "standards of excellence" means writing under the standards of western scholarship that contains and strangles the Filipino language. While I choose to write and teach in Filipino, a language understood by the masses, teaching students unable to grasp academic Filipino forces me to code-switch in English. This strategy of teaching conforms to UP's bilingual policy in education de-emphasizing the use of the national language. Conforming to the global capitalist ideology of the State to use English, UP's fetish for publication in SSCI journals ensures a better position in the university. This means as a Filipino teacher, it is imperative to both publish in English and Filipino. Some of the faculty members despite such predicament, still choose to write *only* in Filipino. Lately, at the University of the Philippines, there is only one journal (indexed in SCOPUS) that accepts Filipino articles. Except for Scopus and SSCI journals, no other journals are recognized as excellent publications.

Today, the Department of Filipino continues to battle the Commission on Higher Education Memorandum Order No. 20, Series of 2013 that eradicates the teaching of 9 Units of Filipino in the General Education (GE) in the tertiary level. The basis of the order is the change in the ten-year basic curriculum in 2011 to a twelve-year basic curriculum known as K to 12 (Kinder to Grade 12) to adhere to international standards. The additional two years would train the student in any of the four tracks she or he will choose whether in academic, sports, arts and design, or technical track. Since most students come from poor families, the emphasis of the state is for students to pursue the non-academic track.

Vocational training is encouraged since it is the more practical track in a globalizing world needing skilled workers abroad. As of now, according to the Philippine Overseas Employment Administration (POEA) website, from 2009 to 2013, the number of workers abroad is 2.2 million. Of course, this does not include the undocumented cases. There are no Filipino courses required in the tertiary level whether in language or literature because the government does not see the need for a higher level Filipino as an academic discipline. As a result, many teachers teaching Filipino are being laid off from different universities. The global capital needs skilled workers and not a university diploma.

The changes in the K to 12 curriculum and the GE curriculum in higher educational institutions are in accordance with the globalization program of the Philippine government. Labor export policy, privatization of government owned corporations and institutions, deregulation, and liberalization in which foreign products and industries put local industries in jeopardy all contributed to the factors that push Filipinos to learn English in order to work abroad. It follows that there is the need to learn and use English as a medium of instruction in basic and higher education. There is no vision to have an industrialized nation because the Philippines is catering to the demands of globalization. There is no need for a higher education in Filipino that can strengthen Philippine studies discourse on identity, nation, and national industrialization.

Being a product of a colonially designed education and a nation that is always undergoing political and economic unrest, I struggle with my love-and-hate relationship with English. While "English for global competitiveness" threatens my security as a Filipino teacher, my passion for literatures written in English cannot be denied. As a fictionist and teacher in Filipino, I am indebted to theorists/critics of children's literature in English and authors of children's fantasy literature from Tolkien to JK Rowling, from Pullman to Gaiman.

The so-called magic realists Esquivel, Allende, and especially Marquez invade my thoughts while teaching the craft of creative writing in Filipino. Filipinos have an affinity to these writers because they are writing the history and pain of those of us living in the "Third World." "Magic realism" is a nomenclature mapping that marginalizes anything that the West does not understand. Philippine culture believes in elementals, in vampiric beings called *aswang*, in talking to the dead, in *kulam* (witchcraft), amidst political corruption, demonstrations, and salvaging. What the West calls magic is a part of the everyday life of Filipinos. This is one way to combat imperialism; to teach reflecting the point of view of the voices suppressed by colonialism and neo-colonialism. The Latin American writers write from their point of view, from the margins of the people neglected and oppressed. Teaching literature means teaching from one's poverty-stricken context. It is to teach literature that reflects Philippine society with the hope of instilling a nationalist education that would alleviate poverty in the future.

Reflecting upon my subject positions as the Other Other, a "postcolonial" (nothing is really "post" in the term postcolonial) Filipina teaching not in English but Filipino, English has become a "felt" language, surprisingly speaking to my being. With executive orders and memorandums imposing the use of English as a medium of instruction, the question "Do I fear English?" arises in me. The answer is a resounding "No," because in the process of abrogation, I am owning my brand of Filipino English that bespeaks my ethos.

Patricia Melendrez Cruz, an English major who chose to teach in Filipino, stresses in her article "Ingles: Daluyan ng Diwang Pilipino? Isang Pagtaya" ("English: Medium of Filipino Consciousness? An Evaluation") what must be the first duty of a writer: "*ang unang hakbang hakbang ay ang kilalanin ng ating manunulat ang kanyang pagka-Pilipino at mula sa kaalamang ito, sumulat nang mahusay*" (61). "The first step is for our writers to know their identity as Filipinos and from that knowledge, write well." And this is how I teach literature whether it is written in English or Filipino. Students must see that we come from a history of different colonizations, of a violent past that still haunts us as a nation, and we are still struggling to fight the neo-colonial mind.

Historically, English has been taught since the beginning of the twentieth century, as a result, educated Filipinos speak and write the language, as such, it is impossible to not use it for the benefit of one's own culture. Like colonialism, it cannot be erased or dismissed specially in today's globalizing world. Even Constantino, in his seminal essay *The Miseducation of the Filipinos*, used English to combat the neo-colonial education. One must own the language without thinking about the rules of the West. Once English is owned and Philippine English is recognized not by other cultures but by one's own culture, only then can one see the power of owning the language. It can then be a tool for transformation.

English is here to stay, but it does not mean that one can't conquer and own the language and use it as a weapon against neo-colonialism.

NOTES

1. Lumbera, Bienvenido. "To a Young Man Who Lost His Mind." An anthology of poems 1965–1974. Ed. Emmanuel Torres. Manila: Bureau of National and Foreign Information, Department of Public Information 1975. Print.

2. Lumbera, Bienvenido. "Pedagogy." Brown River, White Ocean: an Anthology of Twentieth-Century Philippine Literature in English. Ed. Luis H. Francia. New Brunswick, N.J.: Rutgers University Press, 1993. Print.

3. The essay "Why I Stopped Writing Poetry in English" by Jose F. Lacaba was first published in Seoul Korea in *ASIA: Magazine of Asian Literature*. Volume 2. Number 3, (2007): 35–44.

4. Villa, Jose Garcia. "Lyrics 15," in *Selected Poems and New*. The Bookmark Inc., © 1993,12–13.

5. Prometheus Unbound. Reprinted from *Edad Medya*. Pasig City, Philippines: Anvil Publishing, 2000. First published in *Focus* magazine, July 14, 1973.

6. Salanga, Alfrredo Navarro. "For Edwin Thumboo and All of Us Who Suffer Through English in Asia." Turtle Voices in Uncertain Weather: Poems, 1980–1988. Eds. Virgilio S. Almario. Manila: Cultural Center of the Philippines, 1989. 54–55. Print.

7. Salanga, Alfrredo Navarro."A Philippine History Lesson." Vestiges of War: the Philippine-American War and the Aftermath of an Imperial Dream, 1899–1999. Eds. Angel Velasco Shaw and Luis H. Francia. New York: New York University Press, 2002. Print.

REFERENCES

Almario, Virgilio. *Isang Sariling panahon: Mga Tanaga, Diyona, Dalit, at Tatlong Talingdaw: A Time of One's Own: Poems in the Traditional Short Filipino Forms of Tanaga, Diyona, Dalit, and Three Talingdaw*. Translated by Marne L. Kilates. Manila: National Commission for Culture and the Arts, 2008. Print.

Atwood, Margaret. *The Handmaid's Tale*. New York: Seal Books, 1998. Print.

Bautista, Maria Lourdes S. ed. *English Is an Asian Language: The Philippine Context: Proceedings of the Conference Held in Manila on August 2–3, 1996*. Sydney: Macquarie Library, 1997. Print.

Bloom, Harold ed. *Bloom's Guide: Margaret Atwood's The Handmaid's Tale*. New York: Chelsea House, 2004. 13–14. Web. 31 August 2014.

Constantino, Pamela. Personal Interview. 27 August 2014.

Constantino, Renato. *The Miseducation of the Filipino*. Quezon City: Foundation for Nationalist Studies, 1982. Print.

Cruz, Patricia Melendrez. "Ingles Daluyan ng Diwang Pilipino? Isang Pagtaya." *Patricia Melendrez-Cruz: Filipinong Pananaw sa Wika, Panitikan at Kultura*. Edited by Laura L. Samson, Ruby G. Alcantara, Monico M. Atienza, Nilo S. Ocampo. Quezon City: CSSP and University of the Philippines, 1994. 41–65. Print.

Gaerlan, Barbara S. "The Politics of Pedagogy of Language Use at the University of the Philippines: The History of English as the Medium of Instruction and the Challenge Mounted by Filipino." Dissertation University of California Los Angeles, 1998. Print.

Gonzalez, Andrew. "The History of English in the Philippines." *English Is an Asian Language: The Philippine Context: Proceedings of the Conference Held in Manila on August 2–3, 1996*. Ed. Maria Lourdes S. Bautista. Sydney: Macquarie Library, 1997. 25–40. Print.

Hau, Carolyn. "*Dogeaters*, Postmodernism and the 'Worlding' of the Philippines." *Philippine Postcolonial Studies*. Edited by Cristina Pantoja-Hidalgo and Priscelina Patajo-Legasto. Quezon City: University of the Philippines Press, 2004. 116–130. Print.

Jose, F. Sionil. *Mass*. Manila: Solidaridad Publishing House, 1983. Print.

———. *My Brother, My Executioner*. Quezon City: New Day, 1979. Print.

———. *Po-on*. Manila: Solidaridad Publishing House, 1984. Print.

———. *The Pretenders*. Manila: Solidaridad Publishing House, 1962. Print.

Lacaba, Emmanuel. *Salvaged Prose*. Manila: Miriam and Emanwelga Lacaba, 1992.

Lacaba, Jose. "Why I Stopped Writing Poetry in English." Web blog post. Kapetesapatalism .blogspot.com. 20 July 2008. Web. 10 June 2014.

Llamzon, Teodoro A. *Standard Filipino English*. Manila: Ateneo University Press, 1969. Print.

Lumbera, Bienvenido. "Pedagogy." *Brown River, White Ocean: An Anthology of Twentieth-Century Philippine Literature in English*. Edited by Luis H. Francia. New Brunswick, NJ: Rutgers University Press, 1993. Print.

———. "To a Young Man Who Lost His Mind." *An Anthology of Poems 1965–1974*. Edited by Emmanuel Torres. Manila: Bureau of National and Foreign Information, Department of Public Information 1975. Print.

Mangahas, Rogelio. *Gagamba sa Uhay: Kalipunan ng mga Haiku*. Quezon City: C & E Publishing, 2006. Print.

McFarland, Curtis D. "Subgroupings and Number of the Philippine Languages or How Many Philippine Languages Are There?" *Philippine Journal of Linguistics*. 25. 1 and 2, (June and December 1994). 75–84. Print.

Ong, Elizabeth P. *Equipments, Pictorials and For a While: 284 Common Mistakes in English Made by Pinoys*. Quezon City: New Day, 2006. Print.

Philippine Government. *The Commonwealth of the Philippines*. Manila: Department of Science and Technology–Advanced Science and Technology Institute (DOST-ASTI) and with the National Computer Center (NCC), n.d. Web. *Official Gazette*. 4 July 2015.

Philippine Overseas Employment Administration (POEA). *Compendium of OFW Statistics 2013*. Manila: Department of Labor and Employment. N.d. Web. 7 July 2015.

Ramon Magsaysay Award Foundation. "Bienvenido Lumbera: Citation." 2012. Web. 30 August 2014.

Roy, Arundathi. *Diyos ng Maliliit na Bagay*. Translated by Monico Atienza. Quezon City: Sentro ng Wikang Filipino, Unibersidad ng Pilipinas, 2002. Print.

Salanga, Alfrredo Navarro. *The Birthing of Hannibal Valdez*. Quezon City: New Day, 1984. Print.

———. "For Edwin Thumboo and All of Us Who Suffer through English in Asia." *Turtle Voices in Uncertain Weather: Poems, 1980–1988*. Edited by Virgilio S. Almario. Manila: Cultural Center of the Philippines, 1989. 54–55. Print.

———. "A Philippine History Lesson." *Vestiges of War: The Philippine-American War and the Aftermath of an Imperial Dream, 1899–1999*. Edited by Angel Velasco Shaw and Luis H. Francia. New York: New York University Press, 2002. Print.

———. *Writings in Protest 1972–1985*. Quezon City: Ateneo de Manila University Press, 1993. Print.

San Juan, Jr., E. "Filipino Writing in the United States: Reclaiming Whose America?" *The Likhaan Book of Philippine Criticism (1992–1997)*. Edited by J. Neil Garcia. Quezon City: Likhaan and University of the Philippines Press, 2000. 441–464. Print.

Shegina, Toni Rose Ortiz Luis. "The Language Controversy and Its Dilemma in the Philippines: a Consequential Result of the Evolution of English." Ann Arbor: University Microfilms International Dissertation Services, 1998. Print.

Sison, Jose Maria. "Kabataang Makabayan and its Relevance Today: Message to the League of Filipino Students and the Nagkahiusang Kusog sa Estudyante." *Jose Ma. Sison.* 29 January 2013. Web. 30 July 2014.

Teodoro, Noel V. "The Department of Filipino and Philippine Literature: 1966–1981." N.d. TS. Collection of the Department of Filipino and Philippine Literature. Print.

Torralba, Elyrah Salanga. Personal interview. 8 August 2014.

Villa, Jose Garcia. *Poems 55: The Best Poems of Jose Garcia Villa as Chosen by Himself.* Manila: The Bookmark, 1962. Print.

———. *Selected Poems and New.* Manila: The Bookmark, 1993. Print.

3

Using Chopsticks to Read Knife 'n' Fork English

Teaching American Literature in Taiwan

E-chou Wu

SAVORING ENGLISH, SAVORING INTERACTION

English, in Taiwan, has long been regarded as a subject or major à la mode because graduates of the English Department tended to have relatively modern ideas, were well aware of international affairs, and finely tuned to Western mannerisms. With this in mind, some people felt the English language truly uplifted them over provincialism as it enabled them to transcend their cultural limits. English departments in Asia have thus become more conducive to bridging the differences that exist in the world community.

Language is indeed very alluring, giving utmost freedom and allowing readers to travel back and forth, through time and space. Literature, a distinct art form, is a world created by language, and it is a world of languages. How a literary world of one language is carried to a different world of another is sophisticated because literature is always socially engaged. As we all know, language gives us identity; yet it simultaneously separates us as well. Race, gender, class, religion, love, hatred, joy, or sadness represented in the world of language is to be translated and interpreted into different cultural contexts along with "communicative mutations" in the English departments. These communicative mutations have attracted attention from scholars of literary, cultural, and intercultural studies, who care about what happens to the reception of a deterritorialized text and its intercultural connectivity.

People rely on their experience, direct or inherited, to interpret what's new to them. Julia Kristeva's coinage of intertextuality is based on any spoken or written language having to be related to its linguistic precursor because in signifying systems narratives or texts refer to preexisting texts. Ellen T. Armour puts intertextuality in a nutshell: "[t]exts are not written or read outside the contexts set for them by other texts . . . texts are always already inscribed by other texts; the very condition of possibility of reading and writing is intertextuality. In a certain sense, all reading and all writing doubles readings and writings that have gone before. That doubling is constitutive of reading and writing" (66). The intertextual approach apparently resonates with interpretive learning in which meaning is "constructed," not "discovered." People of different places and generations bring out meanings of a particular object in different ways (Crotty 9). Intertextuality has implications for foreign language education, too. Mary Louise Pratt is appealing for multilingualism as one of the major aims for humanities education in the United States believing it is a difference that makes students learn more profoundly. She thinks the central task of educators of higher learning is "making the familiar strange and the strange familiar" (1284). In both intertextual dialogism and interpretivism, meanings are constructed according to the receiver, whose interpretation is based upon his experience. The receiver would allude to any books he read or any happenings he underwent to determine a "true" meaning of his own. Meaning thus accretes, given that cultural diversity in today's world is increasing. Pratt's cri de cœur for an establishment of multilingual education from elementary to higher levels is to inspire students to learn things radically different from their previous experience, so that learning becomes more thoughtful and intelligent.

INGESTING ENGLISH, INSPIRING TAIWAN

Literature taught in foreign lands produces a dialogue that generates more meaning from the reader's cultural awareness. American literature is an important course in Taiwan's English departments because it is considered a conduit for understanding the United States as an example of the Western world. Prominent roles in Taiwanese society: opinion leader, academic elite, politician, tycoon, artist, scientist—most of them are radiating an American influence in their own domains because individuals who have these roles have received higher education and training from American universities.

The development of English education in Taiwan can be divided into at least four main historical periods: (1) 1928–1947, a period of Japanese occupation, during which the Japanese and Western literatures were taught as humanities education; (2) 1947–1980, when Taiwan received economic and military aid enormously from the United States and duplicated the subject curriculum of American universities; (3) 1980–2009, a time when vast numbers of departments of Applied English were established to teach the practical English needed to supply industries with language manpower; and (4) 2009 to the present, a time in which the English departments are being Asianized and attempting to catch the tide of globalization (Lai). Japanese, American, and global influences, mixed up with Chinese and Taiwanese inherited culture, have made Taiwanese students, who have rarely read anything in American literature and culture before they enter college, apt to provide extremely varied interpretations of literary works.

Eating utensils, chopsticks and knives and forks, are shaped as apt cultural metaphors for the interaction between Taiwanese readers and American literary texts;

hence, they are used here to explore the interpretative relationships between them. At Taiwanese dinner tables, chopsticks handle all the dishes, from meat, fish, and vegetables, to rice, noodles, and peanut kernels, showing a compromising nature to different circumstances, while knives and forks exude a stark antagonistic quality for approaching distinctions in taste. The use of chopsticks discloses the Chinese culture of accommodation. In *Empire of Signs*, Roland Barthes deciphers the code of the chopsticks, whose metaphorical sense could be extended to feature a relationship between the two signs of Eastern readers and Western texts.

> The harmony between Oriental food and chopsticks cannot be merely functional, instrumental; the foodstuffs are cut up so they can be grasped by the sticks, but also the chopsticks exist because the foodstuffs are cut into small pieces. . . . [A] chopstick . . . has a deictic function: it points to the food, designates the fragment, brings into existence by the very gesture of choice. . . . [T]he chopstick, designating what it selects (and thus selecting there and then this and not that), introduces into the use of food not an order but a caprice. . . . Another function of the two chopsticks together, that of pinching the fragment of food (and no longer of piercing it, as our forks do); to pinch, moreover, is too strong a word, too aggressive (the word of sly little girls, of surgeons, of seamstresses, of sensitive natures); for the foodstuff never undergoes a pressure greater than is precisely necessary to raise and carry it; in the gesture of chopsticks . . . there is something maternal, the same precisely measured care taken in moving a child . . . the cook's long chopsticks, which serve not for eating but for preparing foodstuffs: the instrument never pierces, cuts, or slits, never wounds but only selects, turns, shifts. . . . For the chopsticks (third function), in order to divide, must separate, part, peck, instead of cutting and piercing, in the manner of our implements; they never violate the foodstuff: either they gradually unravel it (in the case of vegetables) or else prod it into separate pieces (in the case of fish, eels), thereby rediscovering the natural fissures of the substance . . . the chopsticks transfer the food. . . . [C]hopsticks are the converse of our knife (and of its predatory substitute, the fork): they are the alimentary instrument which refuses to cut, to pierce, to mutilate, to trip . . . by chopsticks, food becomes no longer a prey to which one does violence (meat, flesh over which one does battle), but a substance harmoniously transferred. (15–18)

The accommodating chopsticks, echoing Walt Whitman's "I am large, I contain multitudes," are able to skillfully deal with American literary texts. In a foreign literature class, it's impossible for the Taiwanese instructor to provide a complete literary and cultural landscape for students who lack the least idea what the foreign country is. Instructors bring literary works "into existence by the very gesture of choice" without "an order, but a caprice." Placing various interpretations on the works, Taiwanese readers "pinch," not "pierce," them. Readers give maternal care, "raising and carrying" the texts, not pressing them hard. Moreover, like the cook using long chopsticks, the instructor and the reader "never violate," but give "alimentary" interpretation of the works. It is because of chopsticks that American literary texts won't be "cut, pierced, mutilated, and tripped," and neither will they be treated as "prey." They are "a substance harmoniously transferred."

As a settler and colonialist society Taiwan has integrated its cuisine with as much food as it can get from every part of Mainland China because of high rates of mobility caused by the Civil War 1945–1949, between the Guomindang and the Chinese Communist Party. New immigrants from China brought with them cooking experience and original foods to be blended with the indigenous foodstuff of Taiwan. Therefore there would have been multiple food combinations as well as new revisions of their recipes. This faraway recipe offers another way of eating in a different place. It is a good example of the postmodern conception of "hybridity, re-appropriation, and adjustment" to a new reality (Liao 66). Reading American literature in a Taiwanese classroom catches the same nuance.

AMERICAN TEXTS, TAIWAN REALITY

It seems preposterous for a foreign reader to think of his literary interpretation as fallacious when it is inconsistent with one from a native reader. When American literature travels around Asia, reader response theorists would suggest each interpretative community will put its own viewpoints onto the works and "fill in the gaps" in its own way because any text has diverse realizations and "no reading can ever exhaust the full potential" (Iser 280). This "dynamics of reading," in the words of Wolfgang Iser, explains a multifaceted analysis of literary works in a Taiwanese context. In the era of globalization, a text traveling to or flowing into different parts of the world encounters different historical, political, social, economic, and cultural struggles. An American literary studies course taught in Taiwan in connection with its reception and/or the instructor's effort is to some extent expedient but must be qualified to take into account the position, and circumstances, as well as the mediation and understanding of the learner. The questions of reception and approaches that have been measured are (1) functionalism: learning English through literature; (2) affective approach: from knife-fork English to chopsticks English in which Asian American or Chinese American literature and writings involve alleged Chineseness; (3) parallel approach: identity politics, immigrant and racial experiences helping students imagine their forefathers crossing the Taiwan Strait; and (4) allegorical approach: postcoloniality, the War of Independence, literary regionalism, and local color literature that Taiwanese students are not unfamiliar with.

FUNCTIONALISM

A good many English majors, fascinated with the English language rather than its cultural and literary formations, do not show interest in textual contents, but give their attention to acquiring language skills. They enroll as English majors because they have had a penchant for learning the English language since they were high school or junior high school students. Literary works, in their eyes, are functional, covering all the most practical aspects of the language. Their functional attitude toward literature subverts the traditional purpose of literary education, but it is a broader definition than simply being able to read and analyze literature. They read more of the historical and biographical introductions in the anthology rather than the work itself since the English in the author's biographical sketch is contemporary and usable. Intellectual and cultural backgrounds, as "side dishes" in the anthology, have apparently turned out to be the main dish for these particular readers of the chopsticks community. In this line of reasoning, works by literary stylists, such as Edgar Allan Poe, Gertrude Stein, Vladimir Nabokov, Norman Mailer, and John Updike, are not recommended for students whose primary aim is to learn English.

Below is an excerpt that may serve as an example from the introduction of Mark Twain in *The Norton Anthology of American Literature* that I encourage students of functionalism to learn.

> In any event, Huck Finn has enjoyed extraordinary popularity since its publication more than one hundred years ago. Its unpretentious, colloquial, yet poetic style, its wide-ranging humor, its embodiment of the enduring and widely shared dream of innocence and freedom, and its recording of a vanished way of life in the pre-Civil War Mississippi Valley have instructed and moved people of all ages and conditions all over the world. (Baym 2: 19)

For students who do not have any idea about Mark Twain's character of Huck, a nearly illiterate teenage kid, raised in a white lower-class family, they would be confused by Huck's words and would find it hard to believe that the national father of American literature writes as below:

> I had been to school most all the time, and could spell, and read, and write just a little, and could say the multipliation table up to six times seven is thirty-five, and I don't reckon I could ever get any further than that if I was to live forever. I don't take no stock in mathematics, anyway. (Clemens 2: 37; ch. 4)

Gertrude Stein's experimental technique in *The Making of Americans* may also be used to advise students not to imitate the style, but to understand the deeper layers of meaning behind the language.

> Mostly every one loves some one's repeating. Mostly every one, then, comes to know then the being of some one by loving the repeating in them, the repeating coming out of them. There are some who love everybody's repeating, this is now a description of such loving in one.
>
> Mostly every one loves some one's repeating. Every one always is repeating the whole of them. This is now a history of getting completed understanding by loving repeating in every one the repeating that always is coming out of them as a complete history of them. This is now a description of learning to listen to all repeating that every one is always making of the whole of them. (Stein 2:1098)

"Repeating" might refer to the forming of an ideology, or a national identity, or a social custom, or a tradition shared by a group of people who coexist harmoniously and trust one another. Truly this writing style is not a very apt model for students to imitate.

AFFECTIVE INFLUENCE

To affect or to be affected has a strong tie to emotion or the sciences of emotion. An affective result that is positive and beneficial will encourage "approach to similar stimuli in the future" but a negative one "promotes avoidance." People develop their own personal preferences and dodge their own personal aversions by dint of "experiencing the world and conducting an affective appraisal of these experiences" (Garrett and Young 210). Scholars have been diversifying the affect theory over the past decades as it is being used in a host of fields of study (Leys 434–35). While the relationships between affect and emotion and whether basic emotions stem from cognition or non-cognition (or "non-signifying") are not a major concern of this article,[1] learning foreign languages through the affective approach is one of the central themes in language acquisition. According to Norma Garnett, the hierarchical categories of the affective theory applied to the FLES program (foreign language in the elementary school) are Receiving, Responding, Valuing, Organizing, Characterizing by a Value—all of which are behavioral objectives, and the motivation is the key attitude (375). These behaviors, based upon affective awareness, get learners motivated to desire to read; of course textual influence, especially their familiarity, plays a major role in promoting interest, preference.

To many Chinese readers English written by Chinese Americans is more unambiguous than that by Euro-American authors. From multifarious English expressions, Chinese American writing, be its style or sentence structure, caters to Taiwanese students who are in search of their linguistic identity for English (what kind of English they would like to emulate). Imported blends of Chinese things with American culture and the English language stimulate affective and attitudinal effects: the content is, for one reason or another, similar but not exactly the same as theirs, arousing an appetite for a foreign flavor in Chinese cultural food. Similarly, it is understandable that Chinese love to pay a visit to the Chinatowns in different parts of the world, not only out of nostalgia but also an affective or emotional attachment.

Students are attracted to the appropriation, borrowing, adaptation, rewriting, or fakeness of ancient Chinese myths, legends, and classic tales that they have been familiar with since childhood. Works containing Chinese things by authors such as Jade Snow Wong, Frank Chin, Maxine Hong Kingston, Amy Tan, Gish Jen, David Henry Hwang, and among others, have long been on the reading list. The embodiment of the American Dream, for example, is portrayed in Gish Jen's *Typical American* as the Changs move into their new house. One of the many versions of the American Dream is a decent home, a car, and enough money. The fact that the Changs own a real house with a yard in the suburbs is an exuberant happiness, but the author particularly brings in a comparison of Chinese sentiments, which truly tugs at Taiwanese students' heartstrings.

> The Changs agree—a lawn like this was more than just nature, just life. A lawn like this was America. It was the great blue American sky, beguiling the grass upward. It was the soil, so fresh, so robust, so much better quality than Chinese soil; Chinese soil having been prevailed upon for too many thousands of years. The blades were a bit skinny now, but they would fatten. Of course! After all, this was top-quality grass, grown out of top-quality soil. (159)

The combination of sound into syllables, namely, Chinese phonetic writing, also is an alluring charm of reading, providing a scenario like that Chinese audiences find in the language of some Chinese characters in Hollywood movies. Jen is adroit at designing Chinese phonetic writing along with its English translation, which, in a bilingual sense, helps both English and Chinese readers learn the two languages. Some of the many emotion-related quotes that affect, to some degree, students' desire of language alternation are as follows.

> "'You've gotten completely hutu'—muddle-headed" (16);
>
> "da cai – big vegetables" (62);
>
> "Chi de ku zhong ku, fang wei ren shang ren—eat the bitterest of the bitter, become the highest of the high" (288).
>
> "What does Daddy mean, san banfa?" Callie asked finally.
> "Xiang banfa. Find a way," Helen explained.
> "That's what Chinese people like to say. We have to find a way."
> "Find a way to what?"
> "Typical expression," said Ralph absentmindedly, and went back to talking to Helen in Chinese. (245)

In addition to the archetypical Fa Mu Lan in Kingston's *The Woman Warrior*, or the talk-story genre of *Flowers in the Mirror* and *China Men*, or the folk tales in Tan's *The Kitchen God's Wife*, much of knowledge and subject matter concerning Chinese is highly accepted, making students read back the original texts. Ezra Pound's "The River-Merchant's Wife: A Letter," an adaptation from Li Po's *"Zhanggan Xing" (長干行)*, favorably meets the expectations of students. Two Chinese prosaic day-to-day idioms about puppy love, *"qing mei zhu ma" (青梅竹馬)*—childhood playmates and *"liang xiao wu cai" (兩小無猜)*—being innocent playmates, originating from this poem immediately catch the student's eye and evoke response, despite its inaccurate understanding of the original text.

> You came by on bamboo stilts, playing horse,
> You walked about my seat, playing with blue plums.
> And we went on living in the village of Chokan:
> Two small people, without dislike or suspicion. (2:1236–1237)

It doesn't matter whether Ezra Pound adapted the poem faithfully or unfaithfully from his reading of a research paper on Rihaku (Li Po's name in Japanese) before he rewrote "The River-Merchant's Wife: A Letter;" affected students will study the meanings of intercultural translation through Pound's works.

PARALLEL STUDY

Comparing American national experience to that of Taiwan is epitomized in comparisons in food, for example, "dumpling" with "shuijiao," "bread" with "steam bread," "spaghetti" with "noodle," "pizza" with "scallion pancake," among others, providing a postulation that similar foodstuffs, though different, could be fine-tuned for an understanding of each other's tastes. A parallel study of historical, social, political, and cultural contexts of American literature helps students look at the resemblances and differences between the two societies.

There is an improved trend in teaching foreign languages promoted by ACTFL (American Council on the Teaching of Foreign Languages): instead of focusing on the traditional language proficiencies, attention is devoted to the five "Cs" as goals for learning languages: Communication, Culture, Connections, Comparisons, and Communities (Reeser 772). Culture as a domain for students to learn is of course about how students translate from the American-based source culture into their target culture understanding. The underlying concept of the five Cs, though applicable to language classrooms, is the structuring of learning foreign languages and cultures into a functional and cognitive module that will integrate students' native cultural competency into the foreign one rather than keep aloof from it. While teaching cross-cultural literacy, the first task instructors take on is "problem-solving" and second, "imparting knowledge" (Taillefer 505). Accordingly, it calls into play what students do while reading their Chinese texts and what they will do with American texts in their country, Taiwan. Consequently, when reading William Bradford's *Of Plymouth Plantation*, it is quite logical to suggest students imagine their forefathers crossing the Taiwan Strait from Mainland China and confronting the fierce north-easterly monsoon in the seventeenth century. Arriving on the island, facing the wildness, encountering indigenous Taiwanese, missing friends and relatives on the other different shores, the early Minnan migrants did experience what Bradford wrote in *Of Plymouth Plantation*:

> Besides, what could they see but a hideous and desolate wilderness, full of wild beasts and wild men—and what multitudes there might be of them they knew not. . . . For summer being done, all things stand upon them with a weatherbeaten face, and the whole country, full of woods and thickets, represented a wild and savage hue. If they looked behind them, there was the mighty ocean which they had passed and was now as a main bar and gulf to separate them from all the civil parts of the world. (1: 176)

Moreover, Sui Sin Far's (Edith Eaton's) "In the Land of the Free" is a write-back story, in which the whites have become villains. By doing this, she is countering such mainstream authors as Bret Harte, Ambrose Bierce, Frank Norris, and Jack

London, whose Chinese characters are servile, sneaky, or vicious. An example of parallel reading or comparison can be done with Mary E. Wilkins Freeman's "A New England Nun," whose female protagonist Louisa, being chained for fourteen years to a marriage commitment, has been waiting for Joe to come back from his fortune-hunting in Australia. Likewise, a Chinese folk legend, *Xue Pinggui* and *Wang Baochuan (薛平貴與王寶釧)*, that has been adapted into a Taiwanese folk opera, Peking opera, and TV series, and known among Chinese communities, has an eighteen-year waiting for the return of her husband. In spite of the folklore that is about love and war, an aristocratic girl and a peasant boy, issues of the story for discussion are women's social roles, women's lack of economic opportunities and inequalities, women's reactions to emotional events and relationships, and men's myth of women's endless love for them.[2]

ALLEGORICAL READINGS

In family or wedding banquets in the Chinese community, the food name in the dishes is of particular significance. Taiwanese love to use homophones to exhibit an auspicious sign connecting good luck with the food. For example, chicken *(雞)* and house *(家)* pronounced the same as "ke" in Taiwanese, so in the house warming banquet, diners eat chicken signifying a "housewarming" (*Tsia'h ke, ē khí-ke 食雞,會起家*). There are numerous food names used for this purpose, such as fish (*yu* 魚) and surplus (*yu* 餘) making the idiom *nian nian you yu 年年有餘* and meaning there will be a surplus year after year; cake (*gao* 糕) and high (*gao* 高) making *bu bu gao sheng 步步高陞*, rising step by step; Chinese chive (*jiu* 韮) and long (*jiu* 久) making *chang chang jiu jiu 長長久久*, everlasting and unchanging, among others. It is no wonder what is eaten in the mouth means far more than its mere taste because of its sounds and what is spoken through the mouth. This is food mysticism, an allegory of food encounters, an indescribable political and social signification in American literary texts that can be digested by Taiwanese sensibility too.

In the essay, titled "John Bull," from *The Sketch Book of Geoffrey Crayon*, Gent., Washington Irving, among many other authors, agreed that this peculiar character, John Bull, throws light on the Anglo-American relationship on their "current political concerns including burning constitutional questions relevant to revolution, and discussed America's developing independent identity and role in the Atlantic community" (Clark 15–16), especially on how Britain got involved in European politics and American matters. It is through allegorical method, similar to the political cartoonist creation of John Bull, that Irving's contemporaries presented American attitudes toward the undesirable qualities of the English. Putting "Rip Van Winkle" into Taiwanese contexts, more often than not students would paradoxically link their present political situation to the moment when Rip awakes to find he is not an Englishman.

Rip looked, and beheld a precise counterpart of himself, as he went up the mountain—apparently as lazy and certainly as ragged. The poor fellow was now completely confounded. He doubted his own identity, and whether he was himself or another man. In the midst of his bewilderment, the man in the cocked hat demanded who he was, and what was his name.

> "God knows," exclaimed he, at his wit's end; "I'm not myself—I'm somebody else—that's me yonder—no—that's somebody else, got into my shoes—I was myself last night, but I fell asleep on the mountain, and they've changed my gun, and everything changed, and I'm changed, and I can't tell what's my name, or who I am!" (1: 945)

Rip's dilemma echoes Taiwan's. Taiwan is more vulnerable than it seems; its people hold two beliefs in identity politics, independence or unification with Mainland China; China being an economic powerhouse is understandable but the independence issues are always tied to national security and start destroying China-Taiwan relations and objecting to Chinese things. It is the complexity of identity.

The question of identity is frequently addressed to students who read many of the works anthologized; paradoxically, the question remains open. American authors, almost all of whom are settlers or later immigrants, were given one central theme with which to work: What is an American? The dramatic influx of new people into this new land seeking asylum and opportunity from every part of the world serve, in very different ways, to establish the assumptions of different identities: national, racial, political, cultural, religious identities, which a passport is unable to embody. In *Letters from an American Farmer*, J. Hector St. John de Crevecoeur, posed the question of what an American is:

> What then is the American, this new man? He is either a European, or the descendant of a European, hence that strange mixture of blood, which you will find in no other country. I could point out to you a family whose grandfather was an Englishman, whose wife was Dutch, whose son married a French woman, and whose present four sons have now four wives of different nations. He is an American, who leaving behind him all his ancient prejudices and manners, receives new ones from the new mode of life he has embraced, the new government he obeys, and the new rank he holds. He becomes an American by being received in the broad lap of our great Alma Mater. Here individuals of all nations are melted into a new race of men, whose labours and posterity will one day cause great changes in the world. (1: 643)

The work of a powerful writer like Crevecoeur contains a high degree of political content about the construction of identity in an immigrant society. His writing is both highly influential and challenging for Taiwanese readers who are struggling to negotiate their identities.

In addition, Thomas Paine's *The American Crisis* listing the reasons for independence is a compelling and compassionate argument, suggesting for Taiwanese what they could write if there were a war with the other side of the Taiwan Strait. The debates between the Whigs and the Tories are used to compare and contrast with ones held between Taiwan separatists and Chinese reunificationists. Students are galvanized into drafting their own statement for the event of an invasion when they read Paine's frequently quoted words: "It is the madness of folly, to expect mercy from those who have refused to do justice; and even mercy, where conquest is the object, is only a trick of war; the cunning of the fox is as murderous as the violence of the wolf, and we ought to guard equally against both" (1: 704). All this is underpinned by the words of "The Declaration of Independence" for people who need to explain why they must fight.

INTERPRETIVE BORDERS, INTERPRETATIVE IDENTITIES

In "Allegories of Creativity and Territory," Roger Bartra expounds artistic identities creating a debate about whether there is still a cultural territory for literary creations in an era of globalization and the Internet. As he observes,

> [A]rtistic creation emanates from territory defended by ethnic, local, or national groups, which perceive globalization as a threat to their cultural identity. The blurring of boundaries will then provoke a kind of cultural schizophrenia, a loss of the borders of the ethnic and national self that, trapped in a display case of a global museum, will no longer be able to express itself; instead, it will be exposed to the transit of limitless influences and subjected constantly to pillage and prurient stares. (115)

Based upon Gilles Deleuze and Felix Guattari's notion of deterritorialization and reterritorialization, artists and their artistic works are in a fluid state. Deleuze and Guattari use the metaphor for the everlasting shifts, a concept of "'double-becoming,' where one deterritorialized element serves as a new territory for another deterritorialized element" (Holland 59). Deleuze and Guattari say, "Nomadic waves or flows of deterritorialization go from the central layer to the periphery, then from the new center to the new periphery, falling back to the old center and launching forth to the new" (qtd. in Sussman 980). When the American text is reterritorialized and reinterpreted in Taiwan far from its central layer, the reinterpretation becomes new, but in a different time and space this reinterpretation would be deterritorialized. As Roger Batra reiterates, for writers and their works it is because of deterritorialization that "the weakening of meanings is not an elective alternative but rather an inevitable point of departure" (117).

Like the hybrid and relocated authors, the reader's interpretation is in flow too. As an old Chinese saying goes, "*橘逾淮而為枳 ju yu huai er wei zhi*." It means that the orange is changed into a zhi north of Huai River (Zhi is a kind of orange with thick skin and thin pulp).[3] Any cultural product traveling to a new land will be influenced by the new customs and environment and given a new meaning. Relocating the work wherever needed, instructors select apposite texts that resonate with deterritorialization and reterritorialization for discussion and study of intercultural relations ranging across a vast horizon of time and space.

NOTES

1. The debate is thoroughly discussed by Ruth Leys.

2. In an email discussion with Daniel Lehman, who also taught "A White Heron" in his American literature class being a visiting professor at Providence University, he mentioned about half of his students choosing to write a short essay on "A White Heron" in a written quiz. Lehman said,

> [A]ll but one said they believed that Sylvia did the right thing in protecting the heron. Interestingly, by far the majority of them used an environmental claim to back up their position (protecting nature over potential profit). Others primarily used a feminist claim, and a few others made a traditional, rural claim as opposed to the encroaching city life that the bird collector might represent. . . . Only one student thought that Sylvia should have accepted the money on behalf of her grandmother and perhaps explore marriage opportunities with the young man. She wrote: "I personally think that Sylvia should accept the 10 dollars. Even though marriage is very important to most girls, but Sylvia might someday find the good from the hunter and fall in love with him. In Chinese tradition, there were parents who led their children to meet each other one time and get marr(ied) as soon as possible (when) they didn't even have chance to get to know each other, but some of the people kept their marriage relations until now. So I think if you have to do this, then do it." It was no wonder that Lehman was impressed by the responses from students of different cultural background.

3. The translation is from the Chinese-English Dictionary of Modern Usage, edited by Lin Yutang. http://humanum.arts.cuhk.edu.hk/Lexis/Lindict/.

REFERENCES

Armour, Ellen T. *Deconstruction, Feminist Theology, and the Problem of Difference: Subverting the Race/Gender Divide*. Chicago: University of Chicago Press, 1999. Print.

Barthes, Roland. *Empire of Signs*. Trans. Richard Howard. New York: Hill and Wang, 1989. Print.

Bartra, Roger, and Dierdra Reber. "Allegories of Creativity and Territory." Trans. Dierdra Reber. *America: The Idea, the Literature*. Spec. issue of *PMLA* 118.1 (2003): 114–19. Print.

Baym, Nina, et al., eds. *The Norton Anthology of American Literature*. 5th ed. 2 vols. New York: W. W. Norton, 1998. Print.

Bradford, William. *Of Plymouth Plantation*. Baym et al. 1: 165–204.

Clark, Jennifer. "John Bull's American Connection: The Allegorical Interpretation of England and the Anglo-American Relationship" *Huntington Library Quarterly* 53.1 (1990): 15–39. Web. 15 April 2013.

Clemens, Samuel L. *Adventures of Huckleberry Finn*. Baym et al. 2: 28–216.

Crotty, Michael. *The Foundations of Social Research: Meaning and Perspective in the Research Process*. Thousand Oaks, CA: SAGE, 2005. Print.

Freeman, Mary E. Wilkins. "A New England Nun." Baym et al. 2: 560–68. Print.

Garnett, Norma A. "Cognition and Affect for Successful FLES* Programs: Are Both Domains Vital?" *Hispania* 81.2 (1998): 373–78. Web. 15 April 2013.

Garrett, Paula, and Richard F. Young. "Theorizing Affect in Foreign Language Learning: An Analysis of One Learner's Responses to a Communicative Portuguese Course." *The Modern Language Journal* 93.2 (2009): 209–26. Print.

Holland, Eugene W. "Deterritorializing 'Deterritorialization': From the 'Anti-Oedipus' to 'A Thousand Plateaus.'" *Deleuze & Guattari*. Spec. issue of SubStance 20.3(1991): 55–65. Web. 15 April 2013.

Irving, Washington. "Rip Van Winkle." Baym et al. 1: 936–48. Print.

Iser, Wolfgang. *The Implied Reader: Patterns of Communication in Prose Fiction from Bunyan to Beckett*. Baltimore: Johns Hopkins University Press, 1974. Print.

Jen, Gish. *Typical American*. New York: Plume, 1992. Print.

Kingston, Maxine Hong. *The Woman Warrior: Memoirs of a Girlhood Among Ghosts*. New York: Alfred A. Knopf, 1976. Print.

———. *China Men*. New York: Alfred A. Knopf, 1981. Print.

Lai, Chung-hsiung. "Up or Out?: Looking at the 'Transformers 4' of the English Department in Taiwan from an Asian Perspective." National Cheng Kung University. Center for Public and Business Administration Education, National Chengchi University, Taipei. 12 September, 2014. Keynote Address.

Lehman, Daniel W. "Another Note on 'A White Heron.'" Message to E-chou Wu. 24 March 2015. E-mail.

Leys, Ruth. "The Turn to Affect: A Critique." *Critical Inquiry* 37 (2011): 434–72. Web. 7 Feb. 2015.

Li, Ju-chen. *Flowers in the Mirror*. Trans. and ed. Lin Tai-yi. Berkeley: University of California Press, 1965. Print.

Liao, Ping-hui. *Ways of Eating: Savoring Postmodernity*. Taipei: Fish & Fish Publishing, 2004.

Paine, Thomas. "The Crisis, No. 1." Baym et al. 1: 699–705. Print.

Pound, Ezra. "The River-Merchant's Wife: A Letter." Baym et al. 2: 1236–37.

Pratt, Mary Louise. "What's Foreign and What's Familiar?" *PMLA* 117.5 (2002): 1283–87. Print.

Reeser, Todd W. "Teaching French Cultural Analysis: A Dialogic Approach." *The French Review* 76.4 (2003): 772–85. Web. 15 April 2013.

St. John de Crevecoeur, J. Hector. *Letters from an American Farmer*. Baym et al. 1: 641–65.

Stein, Gertrude. "Introduction." *The Making of Americans*. Baym et al. 2: 1093–105.

Sussman, Henry. "Deterritorializing the Text: Flow-Theory and Deconstruction." *Comparative Literature*. Spec. issue of *MLN* 115.5 (2000): 974–96. Print.

Taillefer, Gail F. "Foreign Language Reading and Study Abroad: Cross-Cultural and Cross-Linguistic Questions." *The Modern Language Journal* 89.4 (2005): 503–28. Print.

Tan, Amy. *The Kitchen God's Wife*. New York: Ivy Book. 1991. Print.

II

CULTURE AND LANGUAGE STUDY IN TAIWAN

4

Teaching Early British Literature in Southern Taiwan

Elyssa Y. Cheng

On September 12, 2014, the GEAT (abbreviation for the Global Education Association in Taiwan) held a symposium titled, "Future Directions of the English Departments," and in it, announced the results of their survey, "How to Improve the Administration and Management Environment in Taiwan."[1] Chao-ming Chen, Director of the GEAT and Chair Professor of the Department of Applied Foreign Languages, Shih-Chien University (Taipei campus), proclaimed that out of the 74 valid samples they collected from the 126 questionnaires they distributed to the English departments in Taiwan (including English departments in public and private universities, vocational and technological universities, and five-year colleges), 31.5 percent of English departments in Taiwan are under the pressure of under-recruitment; nearly 40 percent of them have difficulties finding internship opportunities for their students. According to Chen, the survival crisis of the English departments in Taiwan derives from the wide gap between their curriculum and the practical demands of the industries.

Chen pointed out that English majors used to be the pillars of Taiwan's international trade. In the past, English departments were one of the most popular majors in Taiwan. However, as with the increasing demands of specialized knowledge in all fields, non-English majors began to emphasize their students' English training, thus making the English majors lose their prestige in English. Chen asserted, despite the changing economy and global demands for specialized knowledge, English departments in Taiwan still focus their curriculum on English and American literature, linguistics, and English teaching. The limited training in merely English language and literature makes it even more difficult for English majors to find jobs after their graduation.

In the symposium, Chen also pointed out that we should not neglect the gap between the English majors' education and their job placements due to the number of English majors in Taiwan. In the 161 universities and colleges in Taiwan, 126 of them have English or English-related departments. In the 2012 academic year, Taiwan had altogether 51,482 English majors (including 3,432 MA and PhD students), and on the average, each academic year we have approximately 11,500 to 12,000 English majors graduating from their programs.[2] Chen asserted that the significant amount of English majors in Taiwan makes it a pressing issue for us to reconsider the curriculum and future directions of our English departments, and the wide gap between the education our students receive and the practical demands of the industries. Chen noted that most of the English departments in Taiwan aimed to train their students to be future English teachers, secretaries, and literary scholars, but as with the low birth rate, the number of wandering high school teachers and PhD holders rapidly increase. In the future, the English majors shall experience more and more difficulties finding jobs if they are only trained to be English teachers. Based on the statistical data provided by the GEAT, Chen stated that it is time for the English departments in Taiwan to confront their survival crisis. Chen said that from 2007 up to now, the gap between the curriculum and the practical demands of the industries has propelled eight schools to permanently close their English departments and the other nine schools have decided to transform their English departments to departments of tourism and hospitality. At the very end of this symposium, Chen urged that the English departments in Taiwan should reconsider their positions as "breeders of English teachers." They should take advantage of their students' English training and proficiency to cultivate multi-disciplinary talents.

Although, as Daniel J. Bauer notes in his commentary essay "We Must Distinguish between Studying and Learning English" in *The China Post*, the 74 valid samples out of the 126 questionnaires may not absolutely reflect the difficulties that English departments in Taiwan currently encounter, the survival crisis of English departments that Chen singled out not merely affects the lower-ranking private universities and vocational colleges, but also challenges top-ranking universities. The Department of Foreign Languages and Literatures in National Taiwan University (locally called Taida and will be referred to as so later in this essay), supposedly the top English department in Taiwan, also encountered the challenge of changing their curriculum. Tseng Li-ling, the current chair of Taida's English department, stated that the number of graduate entrance exam test-takers were only half of the number of three years ago.[3] Tseng admitted that such drastic dwindling of test-takers was a big setback to professors of English literatures in Taida, who were most proud of their strengths in literary theory and criticism; however, to meet this challenge, the English faculty in Taida were forced to make their compromise and lower their MA graduation threshold by reducing their MA thesis requirement from eighty pages to sixty. In doing so, they expected to let their future MA students finish up their program in two and a half years, which is a year faster than the original. They hope that by lowering their graduation standard, they will be able to attract more students in the future.

In addition to this transformational change, the English faculty in Taida also has readjusted their curriculum by pulling off all the year-long survey courses of English and American literature and turning these courses into semester-long period introduction courses. In the past, Taida's English majors were requested to take three year-long survey courses of English and American literatures, but now they can choose five out of the nine available semester-long courses, and each of them provides students with a more detailed guided reading of a given literary period. Another conspicuous feature of Taida's curriculum change is that they try to divide their courses into groups: Aside from the more conventional literary courses, they now add another group for courses on cultural issues that link traditional literary courses to various sorts of disciplines; another group of courses focuses on translation/interpretation skills that incorporate their newly established translation/interpretation MA program and their conventional linguistics-track courses.

Taida's curriculum change definitely reflects the practical needs of the industries and the global demands of multi-dimensional talents, but it is also a warning to all the instructors of English and American literature survey courses in Taiwan because the curriculum in Taida's English department has always been the guidepost for all English departments in Taiwan. I have been teaching a year-long survey course of early British literature in National University of Kaohsiung (a public university in southern Taiwan) ever since I received my PhD in English from the State University of New York at Buffalo. As an instructor of a survey course, I can see an apparent dwindling of interest in the study of English literature. In Taiwan, most of the instructors (no matter a PhD holder or not) have to teach literature as well as language training courses. Since I have to do both, I can clearly observe my students' reactions in both practical language training courses and literature classes. Most of my students show strong interest in practical language training courses (such as listening, speaking, writing, translation, and even higher-level English courses for special purposes), but many of them express their lack of interest in taking literature courses. In my observation, I also discover that students' learning motivation can definitely affect their performances in class and in tests. For some students, their performances in practical language-training classes and literature courses can be so drastically different that they sound like different persons. At first, like my other colleagues, I blamed their lack of interest on their poor learning attitudes and the changing economic climate, but as I worked hard to design my survey course and applied the teaching techniques that I acquired as an English TA at SUNY and from all the teaching practicum courses that I took (including "Practicum in Teaching" for SUNY's English TAs and Professor Barbara Bono's "Teaching Shakespeare"), I found that the ways an instructor of English literature teaches will greatly affect his or her students' receptions of the class. Therefore, in this essay, I will first explain why I abandon the Asian conventional lecture-based paradigm to adopt a more interactive and activity-based model. Then, I will demonstrate some of the in-class activities I employ to illustrate how these activities help elevate my students' interests in learning early British literature, and I will also briefly introduce my students' receptions to each activity. And finally, I want to reflect upon Chao-ming Chen's argument and Taida's curriculum change to explore the reasons why these courses are not well-received by English majors in Taiwan and ponder about the "keep" or "drop" of these survey courses.

Due to the limitation of educational resources in Taiwan, most English departments can only afford to have small-capacity classes for language proficiency courses, such as listening, speaking, writing, and translation. Survey courses of English and American literature are always taught in large capacity classes of fifty-five to sixty-five students, which can be larger, depending on the number of transfer students and repeaters per year. The size of the class makes it difficult for the instructors to conduct interactive activities, so most of them rely on the lecture format. While lecture makes up the core of the class, I do not think that with lectures alone we can motivate our students and guide them as to how to read archaic English. Before I took teaching practicum related courses in SUNY, the only techniques I knew about teaching English literature were lecture, group presentation, role-play, and film-screening. My English literature teacher used to lecture for three hours, but sitting in class for three hours made me feel so bored that it was really a torture. I hated making and listening to my classmates' group presentations, too, because it took me and my classmates so much time to prepare but did not necessarily guarantee that we understood the content of each other's presentations. Having students make role-plays was fun, but repeating the same activity made me feel that I was enrolled in a drama club, rather than an English department. Watching a film was always better than a horrible lecture, a boring presentation, or a dull role-play, but what was the point of screening a film for three hours straight? To be honest, most of the film adaptations of classical literature were not as attractive as Hollywood movies, and enjoying a film did not necessarily indicate that I enjoyed that literary work or that I understood how to read a literary work critically. When I had a chance to design a survey course on my own, I knew that it was time for me to apply the alternative teaching techniques that I was professionally trained to do in SUNY, and I immediately decided to abandon the Asian traditional lecture-based, instructor-centered paradigm and adopt a more innovative, activity-based model. Instead of letting my students stay put to bear with the large capacity class, I divide my students into six groups and hope to take advantage of class dynamics and peer stimulation to pique their interests.

Same as most English literature professors in Taiwan, I use *The Norton Anthology of English Literature* as my textbook and follow closely on the canon. However, unlike my other colleagues, I do not start immediately with literary works; instead, I focus on the introductory section of each major literary period. I do so because I find it immensely useful to explain the social, historical, and cultural backgrounds to my students first. Because they are not native speakers of English and do not have much knowledge of pre-modern and early modern England, the introduction of social, cultural, and historical backgrounds greatly increases their reading comprehension in early British literature. Another reason for first focusing on the introductory section is that early British texts are highly challenging to even some native speakers, not to mention my students. Regardless of how hard I try, I will not be able to cover all the major texts in a three-hour weekly course. Therefore, I believe familiarizing my students with the historical as well as cultural contexts, teaching them the skills and techniques necessary to analyze the texts, and encouraging them to explore further will have a more significant effect on their grasp of literature than attempting to spoon-feed them as many literary works as possible. So, before I venture into any major literary text of a given period, I usually spend three hours explaining the social, cultural and historical backgrounds, introducing the importance of some of the literary texts, and exploring the literary style, techniques, and terminology of that period. This is probably the most lecture-based section in my course, but I consider such foregrounding important to my students' comprehension of early British literature.

Only after a comprehensive survey of a given period will I start to explore its major literary texts. Although Asian students have been known for their respect for teachers and their obedience in class, most of my students do not seem to follow this tradition. They belong to the very capriciously impatient text-message and iPhone generation, so I try to make my in-class activities as varied as possible in order to keep them alert and encourage their participation. Unlike my other colleagues teaching contemporary and modern texts, I have to start my course from one of the most archaic and difficult texts—*Beowulf*, rather than something fanciful and contemporary. In my university, the starting level of our English majors is about low-intermediate, so I have practical difficulties encouraging my students to read *Beowulf*'s original text straight through. Although the use of study guides, such as CliffsNotes or SparkNotes, has been constantly discouraged here in Taiwan, I find them surprisingly useful to arouse my students' interest in early British literature. In a way, my students are put off by the archaic language in early British literature, but not by the story. Getting them into the stories first definitely helps intensify their motivation. These study guides are handy because the publishers upload them on the web, and my students can easily download them. They spare me a lot of time making and producing handouts.

So, very unconventionally, I try to approach *Beowulf* by starting from the plot summary and character list of these study guides. I actually find my students even have difficulties reading these simplified plots and analyses correctly. Vocabulary, basic historical background training and reading comprehension are their initial three major obstacles to the exploration of early British literature, though I understand that, as a literature professor, I cannot stop at guiding my students to read simplified editions. I need to design innovative methods to encourage my students to read the texts. The method I use for *Beowulf* is group poetry reading. I give each group around 150 lines from the Seamus Heaney translation into modern English of *Beowulf*, ask them to read these lines closely, and ask each group to find a passage of ten to twelve lines from their designated reading to read aloud in class. In asking them to read these ten to twelve lines aloud and in group, I can supervise whether they are doing their close reading correctly and whether they can grasp the rhymes and rhythms of these alliterative verses. Due to the lack of an English-speaking environment, most Taiwanese students can read and write English better than they speak or comprehend it. Some of the flaws they show in writing, such as a lack of sense of style and idiomatic collocation and the negligence of harmonious rhythm, are also due to environmental factors. By guiding them to read *Beowulf* slowly, closely, and aloud, I intend to bring them back to learning English by increasing their sensitivity to sounds, rhymes, and rhythms. I also want to alert them to the issue of style and idiomatic collocation. Yet, most importantly, I want my students to know that in designing their own way of reading *Beowulf* aloud, they are gaining their agency in reading, criticizing and presenting the original text themselves. Through this activity, I hope my students can realize that they should not be indoctrinated by any given interpretations or criticism; instead, as they design the way they present and interpret literature, they will find their own critical perspectives.

To some extent, I repeat and extend this innovative method in my guided reading of *Sir Gawain and the Green Knight*. Again, I use available online study guides to introduce the plot and characters to my students, but I encourage them to explore the three seduction scenes in *Sir Gawain and the Green Knight* closely and critically by setting up comparative and contrasting models in the same scene. I ask two groups to do a close reading of a given seduction scene. One of the group is asked to perform the whole scene in pantomime (with only gestures and background music to help develop the plot), and the other group is asked to explain the whole scene to the class and offer their critique of their classmates' performance. This makes the length of the assigned reading slightly longer than that of *Beowulf*. For this assignment, in the close reading students focus on the anonymous Gawain poet's employment of content words and how the content words help intensify the psychological tension between the seducer (the Hostess of the Castle) and the target to be seduced (Sir Gawain). Meanwhile, the peer review/critique of the close reading serves to illustrate to my students how close reading can affect their interpretation of a given literary text and how their interpretation can consequently affect the presentation of that text. This activity moves toward bridging the gap between literary texts and other forms of art, such as theater, art, and music. In addition, students can realize that with decent training and patience, they can manage to read archaic English literary texts critically and interpret any given English literary text even if they are not native speakers of English.

Aside from poetry reading, theatrical performances, and peer reviews, I also use drawing and painting extensively in my classes. Some of my more conservative colleagues consider this teaching technique a waste of time, and still others criticize this method by saying that this activity excludes those who do not love to paint and draw. However, I find my students enjoy painting and drawing a lot, and those who do not love to paint and draw will automatically find their own role in the group activity. For example, they may choose to read the text closely or make the final oral report. To better incorporate art elements into my classes, I carefully prepare a set of pre-class instructions, stipulating what the students should read, setting up the target they should achieve and regulating the time they can spend for this activity. With careful pre-class preparation and the help of pre-designed handouts, my students can always complete their portraits for characters or comic strips for a given plot or a poem in one to one and a half hours in class. They really enjoy this activity, and their enjoyment can be shown in their "stealing" this idea to complete their assignments for other classes. My initial rational of designing this activity is very simple. My students' reading comprehension of early modern literature is relatively low, and my classes are so large that it is very difficult for me to diagnose whether my students are reading a given literary text correctly or not. Of course, like my own teachers before me, I could quiz them to force them to read, but I do not want to use written tests to deprive my students of the pleasure of reading literature. Their paintings and drawings serve as a visible indicator of what they have and have not understood. For me, this activity is pleasurable to the students, and time-saving to the instructors. My students often take photos of themselves in front of their final products and upload their portraits and comic strips onto their facebook pages and personal websites.

"The General Prologue" in Geoffrey Chaucer's *The Canterbury Tales* is the first literary text for which I employ this painting and drawing technique. I created this method due to my painful days of reading and memorizing the features and personality traits of all the twenty-nine characters in "The General Prologue" when I was an undergraduate student. At that time, I understood that all the features and personality traits of the characters were crucial to my understanding of Chaucer's tales and satires, but the memory chip in my brain was too tiny to accommodate all the data I needed to remember. At that time, I decided to draw portraits for all the characters as I read. They surprisingly came alive in my mind. Today, when I teach Chaucer's "The General Prologue," I first give my students a handout, cataloguing all the personality traits of the characters in the prologue, and then give them the freedom to choose their favorite character and draw a portrait of him or her. My students are always impressed with my giving them the freedom to choose their favorite character, while they unknowingly ignore the fact that they have to finish reading the complete prologue in order to decide which character is their favorite. This portrait drawing exercise does not merely serve as an instrument to encourage my students to read Chaucer; it also helps bridge the link between literature and art.

While the portrait drawing exercise helps build the connection between literature and art, comic strip creation develops and extends my students' close reading sensibility to their logical judgment, critical capability, organizational skill, and aesthetic evaluation. I use this technique to teach Geoffrey Chaucer's "The Wife of Bath's Prologue and Tale" in *The Canterbury Tales*, Edmund Spencer's *The Faerie Queene*, Book I, and some seventeenth-century metaphysical poems. In order to create comic strips for a given literary text, the students must read the literary text closely and decide which episode or scene they are going to paint. Due to time limitations, each group will only produce a short four-block comic strip. To complete this assignment, students have to evaluate, judge, and organize the episodes or scenes they want to depict in their strips. From the first time they create their comic strips for "The Wife of Bath's Prologue and Tale" to the moment when they paint their imaginary scenarios for designated seventeenth-century metaphysical poems, I find that my students' critical, organizational, reading and aesthetic capabilities develop and mature. From an instructor's point of view, this activity provides a bird's-eye view of the students' understanding of the assigned reading. From their painting, I can easily discern whether they misread a certain episode or scene. It is an effective and time-efficient means of teaching large-capacity literature classes in southern Taiwan.

To help my students realize the close relationship between literature and other forms of arts, I alternate among different techniques, such as role play, program writing and design, comparative videographic analysis, costuming, locating background music, and shoebox production to teach medieval and Renaissance dramatic texts. Of course, role play is the most common and easiest among these techniques, but I use it with caution and intend to include some creative elements in this activity. Instead of having one group of students acting out a whole play, I try to divide the play into six parts and ask each group to perform a part of it. Also, I prepare a checklist of important items for my students to know what I expect for this mock-theatrical performance. I hope that, aside from memorizing the lines, my students will also notice the other forms of arts in dramatic performances, such as costuming, makeup, stage design, and background music. In order to build up their reading and critical capabilities, I also request them to give me a short note explaining why they make a certain cut or edit from the original text. My students enjoy performing *Everyman* a lot, and most of the time, I am deeply impressed with the way my southern Taiwanese students portray the medieval allegorical figures.

Although most of my students are very active and enjoy performances, sometimes I may have an exceptionally shy and timid class. Their shyness does not indicate that they are not going to be successful English majors, but it is an indication to me that I have to seek some alternative strategies to teach early modern drama. One of my ways of doing this is to ask my students to write and design a program for the play I teach. I have an impressive accumulation of programs from the theatrical productions that I have been to in the States and England. I give my students some that I have collected as examples and ask them to write and design a program for the play I teach. Some of my very shy students may experience stage fright when facing live stage performances, but I am often shocked by their boldness in the costumes they make for themselves to create their programs. Their costume photos also demonstrate their creativity and ability to understand early modern British drama.

Sometimes, I know I have a group of shy students, and from my experience, I can determine whether the group is the one that would not like to spend so much time writing and designing a program for a play. If so, an alternative technique I use for classes like this is to have the students design a mini-shoebox production of a set for a certain scene in a play I am teaching. I will have a checklist cataloguing all my requirements for this production and give them one or two class periods to finish their productions. I find this activity useful because it actually forces my students to read the whole play carefully, so that they can locate the scene they want to design for their shoebox. I also discover making shoebox productions a very good way to encourage close reading because only after reading a designated play closely can my students decide how to position their characters on the stage set they design. When I first employed this activity, I was reserved and concerned whether it would be boring and take up too much of the class time, but when I saw my students photographing the details of their shoeboxes and excitedly uploading the photos onto their Facebook pages, I realized that they enjoyed it, and over the years, I have become bolder and more courageous with this activity.

Although the average English level of my students in southern Taiwan is relatively low, there are always 10 percent of them interested in English literature. These serious students usually take the lead in reading and interpreting the texts for their more practical-minded students in the drawing and cartoon panels, but they will surely not be completely happy if my class activities go only as far as more popular and less intellectual-oriented designs. Thus, when I write my syllabus, I need to keep these students in mind and try to incorporate some serious and scholarly dialogues and debates into my class.[4] I discovered that my employment of multimedia and comparative videographic analysis, along with my provision of discussion sheets, is very popular among these students. For instance, I teach the concept of the Cult of Queen Elizabeth by starting with a powerpoint slide introduction to the royal portraits of Queen Elizabeth I, and then I screen clips from director Shekhar Kapur's films, *Elizabeth* (1998) and *Elizabeth: The Golden Age* (2011) and the TV series *The Tudors* (2010), so that my students may understand the life and sociopolitical contexts of Queen Elizabeth. After watching some clips from these films, I usually guide my students to read Queen Elizabeth's "Speech to the Troops at Tilbury" to further delve into the Queen's construction of her virginal, iconic image. To elevate my students' critical capability, I do not stop after introducing my students with the Cult of Elizabeth alone; I also want to show them how the Elizabethan idolatry was feared and challenged by her contemporary Englishmen. To do this, I teach Shakespeare's *A Midsummer Night's Dream* and use two different film adaptations (one directed by Andrian Noble and the other directed by Michael Hoffman) and Louis Andrian Montrose's excellent critical essay, "Shaping Fantasies: Figurations of Gender and Power in Elizabethan Culture," to help my students understand the Elizabethan Englishmen's anxiety over a nation state ruled by a female monarch. I prepare an in-class discussion worksheet for the construction of Elizabethan idolatry and the male resistance, and with the aid of the multimedia and the help of Montrose's scholarly essay, my students can clearly position their thoughts and form opinions of how they feel about Elizabethan male anxiety over a female monarch.

I also use videos to teach Christopher Marlowe's *Doctor Faustus*, William Shakespeare's *Twelfth Night*, and Jonathan Swift's *Gulliver's Travels*, but I approach each work differently. I like to use Richard Burton and Nevill Coghill's 1968 version of *Doctor Faustus* as a spring board for me to teach *Doctor Faustus*. Perhaps due to our age difference, I do not find this version ancient and bothersome to me, but my students find the performance disturbing. When I press them for the source of disturbance, they tell me that they cannot stand the special effects of the 1960s. Their disturbance actually serves as a good way for them to discuss the differences between Marlowe's original text and the film adaptation. Aside from Burton and Coghill's film, I also use a short clip from Walt Disney's *The Little Mermaid*, describing how the Little Mermaid, Ariel, signs a commercial bond with the sea witch, Ursula, to trade her voice for human legs and accession to the prince she loves. I use this clip to illustrate the similarities between the financial contract made between Doctor Faustus and Lucifer's messenger, Mephastophilis, and that made between Ariel and the sea witch, Ursula. My students are delighted and entertained by the clip from *The Little Mermaid*, and they are always impressed with the introduction of contemporary popular elements into our class.

Teaching Shakespeare's *Twelfth Night* and Swift's *Gulliver's Travels* is always a pleasure for my students and me. Both texts are less heavy and contain dramatic elements that college youth adore. I also use videos as my guides to teach these texts, but I try to make my teaching of each of them slightly different. I use director Trevor Nunn's film adaptation to teach Shakespeare's *Twelfth Night*. This film is very delightful and entertaining, and the actors' deliveries are clear enough for my southern Taiwanese students to understand. Therefore, this is one of the film texts I can screen confidently in the first two hours of my three-hour class. The characters in the play have distinctive personalities, so I can easily design a personality trait worksheet to help my students develop observational and critical abilities. This personality trait worksheet has provided some feedback that has helped self-reflection. In one year, when my students' level seemed to be lower than that of our former students,' I tended to be harsher and more demanding of the students. When I asked my students to associate the personality traits of a certain character in *Twelfth Night* to people in their surroundings, they compared me to a Puritan Malvolio figure. I learned that I had better show a little more patience and be a little less severe. I think comparing film and literature can serve as a good way for me to boost class dynamics and bridge the gap between me and my students.

I use Charles Sturridge's 1996 TV movie *Gulliver's Travels* as a guide to help my students read Swift's original text. Due to the length of the novel, I usually assign it as a reading for the winter break. At first, I was concerned whether my students had access to the complete TV movie; however, they told me that it was not hard for them to find it in the DVD rentals. For this assignment, I ask them to compare the TV movie and the novel and try to find omissions and alterations made by the director. At first, my students always come out with extremely straightforward simple sentences to summarize a whole scene. To help them delve deeper into the differences, I try to encourage them to probe deeper by listing the details they can observe. After using one of the episodes to demonstrate some possible ways to locate and list the details, I assign each group a specific episode/scene and request my students to observe the details. This exercise allows me to show my students two things: the importance of doing close reading and how different interpretations of a given text can affect a director's decision in omitting or making alterations in his film adaptation.

In designing strategies and developing techniques for teaching early modern literature, I find John Milton to be the most challenging of authors. Milton's texts themselves are difficult, no matter if we are reading his elegies, such as "Lycidas," or his epics, such as *Paradise Lost*. I tried to use poetry reading, flashcard displaying, and blank filling, but none of them appeared to be effective. My way of getting into it, finally, is through line-by-line analysis and interpretation—a technique that seems to be dull and old-fashioned as soon as one hears it, but I try to use it more innovatively. Since it is too hard for my students to read Milton's original text at the start, I again resort to the help of online study guides, such as CliffsNotes or SparkNotes, to let them understand the plot and major characters of *Paradise Lost*, Books VIII to X. And then, I try to select passages from the original texts that carry out a traceable thread of thoughts, such as Milton's idea regarding women or Milton's delineation of Satan or Adam. I ask my students to read these designated passages closely and thoroughly, and then request that they interpret a given passage based on the content words they locate in the passage. I have been employing this exercise for eight years, and I can guarantee that it is difficult for most of the students. I used to give each group three to four passages to work on, but now I will only give them two due to time restraint. When I walk around my class to join in student group discussions, I am reminded how hard it is for Asian English majors to read Milton. I oftentimes find them futilely trying to look up words in their electronic dictionaries and struggling to turn Milton's reversed sentences in his verses into typical English word order. I also find it is highly challenging for my students to make sense of the content words that they locate in a given passage and interpret the passage by themselves. Although Milton may be frustrating, I give them this assignment so that they may discover that no matter how difficult a given literary text is, with patience and time, they are capable of reading Milton critically and enjoyably.

From my experience, it is virtually impossible for me to teach my students how to read Milton simply by requesting them to read the Miltonian texts extensively. This will only discourage them from reading Milton. The only way to help them overcome their fear of Milton is to give them a workable method of approaching his work and building up their confidence. I assess the effectiveness of this method by the answers the students give to my essay questions. It appears to be a good method because, in their answers, the students not only present material from the one or two passages they are assigned to read in class, but they also incorporate material from other passages that I have not assigned them to read. Even if this exercise takes up one or two hours of our class time, I still consider it a better way to teach my students how to think, analyze, and read Milton than to force them to read the Miltonian texts extensively, only to find out they are reading his work in translation!

Like most instructors in Asia, although I consider testing a final resort to teach a class, I cannot deny it is a core part of the curriculum for us, and a necessary evil for me to discuss here. As an instructor of a survey course, I can observe that many of my students are, in fact, more interested in English language skills for practical and vocational purposes. Although I can design innovative in-class activities to motivate these students to actively participate in my class and learn the importance of literature, I still have to prepare testing materials for my students at both ends (for those who want to further their study in English literature and for those who take their course simply because it is mandatory). I do this by writing my tests with different levels of difficulties in mind. For my vocation- and practical-oriented students, I design multiple choice and short answer questions to test their reading comprehension and vocabulary level. For my more scholarly students, I give essay questions to elevate their writing and analytic ability and keep them engaged in critical dialogues and debates.

My experience in the past twelve years tells me that as long as an instructor is enthusiastic about teaching and works hard on course design, he or she will also be able to teach a survey course well. Despite the students' dwindling interest in the study of English literature, the results of my teaching evaluation remain high (usually above 4.5 out of 5 as the full score), and for many times, the evaluation results of this course are even better than some of the light-duty language training courses I teach. I believe my innovative and activity-based course design contributes a lot to the success of my teaching, but as I occasionally overheard scathing grievances on other survey courses from my students, I feel that I have the urge to explore why these courses are not well-received by English majors in Taiwan, and I do not think the reason is merely course design skills. First, I find that a lot of instructors of these survey courses are not sensitive enough to our students' limited vocabularies. In the past, Taiwan's senior high schools used to employ different English textbooks from a variety of publishers. At that time, even if the students had a heavier burden and needed to read more textbooks and memorize more vocabularies, they at least were forced to have a broader bank of English vocabulary. However, when the Ministry of Education (locally abbreviated as the MOE), under the consideration of lessening the burden of senior high school students, stipulated a standardized textbook and regulated the English vocabulary to be tested in the Joint College Entrance Exams within the range of the standardized textbook, this educational policy actually heightened the teaching difficulties for instructors of the English and American literature survey courses in Taiwan. How can you expect a student whose vocabulary is just or even below 4,500 words to read *Beowulf* or *Paradise Lost*? Not to mention the fact that each year only single-digit students out of the one hundred thousand entrance exam test-takers receive full score for their English tests.

Aside from our students' low vocabulary level, I also find that, for the Taiwanese English majors, the more they read does not necessarily guarantee better English. My professors in the States and my American colleagues in Taiwan used to tell me that it does, but my teaching experience tells me that this theory does not apply to the Taiwanese students. If they simply read but are reluctant to memorize new vocabularies, they will not make any improvement. Native speakers of English were born to be exposed to the English-speaking environment, and, even if they do not work hard to memorize, a lot of jargon, special terms, and vocabulary have already been part of their lives. Extensive reading is definitely a good trigger to brush up and refine advanced-level vocabulary for them. However, for the Taiwanese students, who do not have an English-speaking environment, reading far beyond their comprehension level simply frustrates them and even deprives them of their interests in reading English literature—a sad fact that a lot of instructors of survey courses are not empathetic enough to comprehend.

Another reason for the Taiwanese English majors to lose interest in studying English literature, especially in taking year-long survey courses, derives from instructors' lack of enthusiasm. Designing a course and preparing handouts and worksheets take time and patience. Many instructors in Taiwan are simply reluctant to spend so much energy on teaching. Our basic salaries for university professors are relatively low in comparison to other neighborhood countries in Asia, such as Japan, South Korea, Singapore, and Hong Kong. Our government intends to keep our basic salaries low in order to "encourage" us to do more research; but ironically, the Taiwanese government's "good will" results in the university professors' fierce competition in research grants at the cost of sacrificing our students. The situation impacts most strongly on the humanities. Our salary raise and research awards (usually monetary in order to compensate for our low basic salary) are rigidly tied up with our tenure-track screen (usually to be screened five to eight years, depending on the university one's in), our teacher's evaluation (usually to be held every three to five years), the number of government-funded research grants we received, and our promotion; but, as anyone can notice, humanity professors in Taiwan can only make more money by participating in the fierce competition of research. Yet, with the decline of Taiwan economy, the competition becomes fiercer and even more ruthless. The salary gap between the assistant professor and the associate professor level is not big, but the biggest jump in salary that a Taiwanese professor can expect is to be promoted to a full professor. The most tricky and bureaucratic side of our promotion network is that it not only evaluates our performances in research, teaching, and academic as well as community services, but also counts in the number of research grants we receive from the MOST (the abbreviation for the Ministry of Science and Technology, the highest research-granting institution in Taiwan). The MOST accepts research grant proposals every December, making research grant submission by Christmas Eve the biggest annual event for Taiwanese professors. Due to its stringent budget, the MOST, of course, cannot sponsor every proposal, and the consequences are the more and more heated competition in the number of journal articles and book chapters one has, the foreseeable mud-slinging reviews among peers, and the redistribution of one's personal time to meet with higher and tougher research demands. Indeed, how can you expect a professor to teach nine to twelve hours a week (usually three to six courses), to submit a research grant proposal every year and to publish four journal articles or book chapters in every five years? And under such a workload, how can you expect a humanities professor to publish ten essays or something equivalent to this quantity in seven years in order to win his or her promotion? Since the pressure of doing research is intense and excellent teaching hardly gets a Taiwanese professor anywhere, many instructors in Taiwan choose to sacrifice teaching so as to save their energy for research. As we can tell from the above analysis, the misleading research policy made by the MOST sacrifices our students' education under the guise of promoting academic research.

Aside from limited vocabulary level and the MOST's misleading research policy, another critical reason that results in our students' losing interests in taking American and English literature survey courses is that English teaching has ceased to be the ideal job for the English majors after their graduation. As with the declining Taiwan economy, the stagnation of salaries, and the soaring housing prices, the youth in Taiwan all hope to have a more promising job after graduation, and being an English teacher is no longer an ideal job. Taiwan's low birth rate reduces employment opportunities in public and private high schools, and the stagnation of salaries in the past fifteen years have made the privately owned cram schools (locally called as *buxiban*) become sweat factories for those hourly paid and contract-based English teachers. Even if one continues to pursue a higher academic degree after his or her college graduation, the employment opportunities for university and college teaching positions are rare; or, they are even more scarce than seeking high school and cram school teaching jobs. For the sake of their survival, our students have already considered jobs in other fields, such as business, mass media, performing arts, and hospitality/tourism.

Most of our students understand and realize that they have the need to develop and consolidate specialized knowledge in the practical fields in order to meet with the challenges of the job market, but there seems to be a wide gap between our students' realization of social reality and the professors' ideal regarding what an English department should be. Most English departments in Taiwan remain a "Creole" combination of required courses in English and American literature, literary theory, cultural studies, linguistics, and even TESOL; and pathetically, even the recent curriculum in Taida fails to recognize this phenomenon. Unlike most of the English departments in English-speaking countries, Taiwan's English departments have always been a contested battlefield between American and English literary scholars, comparatists, literary theoreticians, cultural studies scholars, linguists, and TESOL specialists—all of them hoping to compete for a seat in the field of English studies here. The result is the monstrous amount of the required and "must-take" courses that our English majors should take. Scholars of different areas and disciplines all assert the importance of their fields to the English studies, without even considering the workload and physical endurance of our students. Fulfilling all the requirements of these courses is challenging and stressful enough. If they are additionally interested in developing and strengthening their knowledge in business, mass media, and other practical fields, they may have to extend their program for one more year. Since the extension will often add financial burden to their families, the students in Taiwan usually choose to remain "un-apprenticed" to the practical trades. Taida's curriculum change is definitely inspiring to the future development of English studies in Taiwan, but it seems to me that the change lags far behind the practical demands of the job market and the industries.

As an instructor of early British literature in southern Taiwan, even if I love the field that I have committed to and even if I badly hope to have a chance to teach my specialty every year, I actually agree with Professor Chao-ming Chen that we need to restructure our English programs by taking the changing economic climate and the students' practical needs into consideration. To raise our English majors' interest in learning English and American literature, we should reconsider the curriculum design of our English programs. For instance, in order to elevate our Freshmen's vocabulary level, we can actually move some of the practical, selective courses, such as "Journalistic English," "Business English," "English for Tourism/Hospitality," and "Introduction to Translation" to the first year curriculum so that our English majors can have a clear overview regarding the future prospect of their career and also have a chance to broaden their English vocabulary. To help our instructors of English and American literature be more dedicated to teaching and course design, the MOE and the MOST should look to a longer prospect when they stipulate any educational and research policies. All relevant officials should understand that policy making should not only be concerned with the short-term results, but also long-term impacts. I think the English professors in Taiwan need to learn to make compromises. English departments should cease to be the pitting zones for various disciplines, and our curriculum design should be based on the interest of our students, rather than out of the consideration of having every professor open one or two courses in his or her professional field. Although we all want to use the classes to brush up on our research, we need to keep in mind that our students may not be as interested in our fields as we are. As instructors, we need to learn to put education prior to our research, and both the government institutions and we ourselves should not discount our role in cultivating future talents. We have to know that we are sacrificing our students if we all jump into the restlessly vicious cycle of research competition. I agree with Taida's curriculum change to make English and American literature courses elective, rather than mandatory, but I do not think we need to drop survey courses completely. Survey courses manage to provide literary, historical, socio-economical and cultural backgrounds to non-native speakers. Without them, we will not be able to trace the historical evolution of English and American thinking. I suggest keeping them, but keeping them in the way that we open these courses for those who are interested in western literary and cultural history. I also believe these courses will not be that difficult to teach as long as the instructors have the mind to renovate course design and adopt more popular activities in their classroom.

NOTES

1. The agenda of this symposium was obtained from Ching-huei Hu's "Over Thirty-percent of English Departments Under Recruitment Pressures" in *Chinatimes*, 12 September 2014; Hsiao-Yun Lin's "The Bleak Future of English Departments: Eight English Departments Closed or Suspended in Eight Years" in *Liberty Times*, 12 September 2014; Hsiao-Yun Lin's "No

Career Future for the English Majors: 17 English Departments Change Names or Closed" in *Liberty Times*, 13 September 2014; Hsiu-Chin Lin's "English Departments Losing Popularity? Eight English Departments Closed or Suspended" in *United Daily News*, 13 September 2014; Chia-Yun Liu's "Thirty Percent of English Departments in Taiwan Are Under-Recruitment" in *Apple Daily*, 12 September 2014; Chia-Yun Liu and Chin-Hung Wong's "Under Recruitment and Hard to Get Internships: Eight English Departments Closed" in *Apple Daily*, 13 September 2014; Wan-Chi Yu's "Ignorance of Practical Training Leading Eight English Departments Closed or Suspended"; and an anonymous reporter's "English Departments Under Crisis and Change: The GEAT Announced its Survey on Improving the Administration and Management Environment of the English Departments in Taiwan." These articles were originally written in Chinese and appeared to be very similar. I presume that these reports were transcribed from the GEAT's official press release for the symposium. The only more substantial English source that I can find regarding this symposium is Daniel J. Bauer's commentary essay, "We Must Distinguish between Studying and Learning English," in *The China Post*, 28 September 2014. Bauer writes to refute Chao-ming Chen's proposal for applied English and emphasizes the importance of reading English literature based on his English-Chinese bilingual background and his experience as an English professor of Fu Jen Catholic University in Taiwan for more than twenty years. I agree with Bauer's idea that American and English literature should remain an essential part of the curriculum, but I think Chen's proposal also makes sense because we definitely have to tend to our students' career future and narrow down the gap between what we teach them in school and what the industries demand.

2. Statistical data of students is gathered from the newspaper articles as referenced in the previous note.

3. Chair Tseng's interview is reported by Hsiu-Chin Lin, released in *United Daily News*, 13 September 2014.

4. Due to the gap between the intellectual and English levels of my students and due to the different needs of my practical-minded students and more serious students, I need to be exceptionally careful when designing my syllabus. I try my best to keep at least one half of the class time for lecture and the rest of the time for in-class activities. From my experience, unlike the conventional rationale that the more intellectual students will hate portrait drawing and comics painting panels, they like them, but I find them more eager to take the lead roles in these panels. They want to dominate the discussion and determine what to draw/paint. I believe there are three major factors that contribute to the success of these in-class activities—the Asian mentality of winning and getting a higher score, the introduction of iPhone and the interactive media age, and the long and extended three-hour class session that makes students eager to use the activity time to relieve their boredom and stretch their legs.

REFERENCES

Anon. "English Departments under Crisis and Change: The GEAT Announced Its Survey on 'Improving the Management Environment of the English Departments in Taiwan.'" *CNA*, 13 September 2014. *CNA* Web. 7 October 2014. http://www.cna.com.tw/postwrite/Detail/155958.aspx#.VGqwZPmUd9x.

Bauer, Daniel J. "We Must Distinguish between Studying and Learning English." *The China Post*, 28 September 2014. *The China Post* Web. 7 October 2014. http://www.chinapost.com.tw/commentary/the-china-post/daniel-j.-bauer/2014/09/28/418173/We-must.htm.

Burton, Richard, and Nevill Goghill, dir. *Doctor Faustus*. Columbia Pictures, 1968. Film.

Clements, Ron, and John Musker, dir. *Little Mermaid*. Walt Disney Pictures, 1989. Film.

Elizabeth I, of England. "Speech to the Troops at Tilbury." *The Norton Anthology of English Literature*. Edited by Stephen Greenblatt. 9th ed. vol. 1. New York: Norton, 2012. 762. Print.

Greenblatt, Stephen, gen. ed. *The Norton Anthology of English Literature*. 9th ed. vol. 1. New York: Norton, 2012. Print.

Hoffman, Michael, dir. *A Midsummer Night's Dream*. Twentieth Century Fox, 1999. Film.

Hu, Ching-Huei. "Over Thirty-percent of English Departments under Recruitment Pressures." *China Times*, 12 September 2014. *China Times* Web. 7 October 2014. http://www.chinatimes.com/realtimenews/20140912003386-260405.

Kapur, Shekhar, dir. *Elizabeth*. Universal Pictures, 1998. Film.

———. *Elizabeth: The Golden Age*. Universal Studios, 2011. Film.

Lin, Hsiao-Yun. "The Bleak Future of English Departments: Eight English Departments Closed or Suspended in Eight Years." *Liberty Times*, 12 September 2014. *Liberty Times Net* Web. 7 October 2014. http://news.ltn.com.tw/news/life/breakingnews/1104490.

———. "No Career Future for the English Majors: 17 English Departments Change Titles or Closed." *Liberty Times*, 13 September 2014. *Liberty Times Net* Web. 7 October 2014. http://news.ltn.com.tw/news/life/paper/812659.

Lin, Hsiu-Chin. "English Departments Losing Popularity? Eight English Departments Closed or Suspended." *United Daily News*, 13 September 2014. *United Daily News* Web. 7 October 2014. http://mag.udn.com/mag/edu/storypage.jsp?f_ART_ID=535178.

———. "Taida's Foreign Language Department: Test-takers for Graduate Admissions Exam Cut into Half." *United Daily News*, 13 September 2014. *United Daily News* Web. 7 October 2014. http://mag.udn.com/mag/edu/storypage.jsp?f_ART_ID=535179&kdid=AR10&r=4.

Liu, Chia-Yun. "Thirty Percent of English Departments in Taiwan Are Under-Recruitment." *Apple Daily*, 12 September 2014. *Apple Daily* Web. 7 October 2014. <http://www.appledaily.com.tw/realtimenews/article/new/20140912/468471/>.

———, and Chih-Hung Wong. "Under Recruitment and Hard to Get Internships: Eight English Departments Closed." *Apple Daily* 13 Sept. 2014. *Apple Daily* Web. 7 Oct. 2014. http://www.appledaily.com.tw/appledaily/article/headline/20140913/36082216/.

Maclean, Alison, Brian Kirk, Charles McDougall, Ciaran Donnelly, and Steve Shill, dir. *The Tudors: The Complete Series*. Showtime Entertainment, 2010. DVD.

Montrose, Louis Andrian. "Shaping Fantasies: Figurations of Gender and Power in Elizabethan Culture." *Representations* 2 (1983): 61–94. Print.

Noble, Andrian, dir. *A Midsummer Night's Dream*. Lionsgate, 2002. Film.

Nunn, Trevor, dir. *Twelfth Night or What You Will*. Image Entertainment, 1996. Film.

Sturridge, Charles, dir. *Gulliver's Travels*. Hallmark Home Entertainment, 1996. DVD.

Yu, Wan-Chi. "Ignorance of Practical Training Made Eight English Departments Closed or Suspended." *United Daily News*, 12 September 2014. *United Daily News* Web. 7 October 2014. http://mag.udn.com/mag/edu/storypage.jsp?f_ART_ID=535091.

5

Teaching English Language and Literature in the Asia-Pacific Region

Environmental and Ecocritical Contexts

Iris Ralph

When I first moved to Taiwan, I took up a teaching position in the English Department in the College of Foreign Languages and Literatures at Tamkang University (TKU), an institute that is internationally recognized for its ecocriticism conferences, undergraduate and graduate courses, guest lectures, and summer seminars (hosted by TKU's English Department). As an English literature scholar and an ecocritic, I was delighted to secure this teaching position as well as to learn that the department chair at the time, Peter I-min Huang, was one of Taiwan's foremost ecocritics, a scholar who was and continues to be highly respected for his efforts in promoting ecocriticism, an area of inquiry that, as it is taught in English Departments, might be called "ecocritical English" and is broadly defined as "an environmental version of English Studies" (Kerridge 13).[1] I looked forward to learning more about the environmental issues that were specific to Taiwan and China, so after I arrived in Taiwan I began to research those issues and bring attention to them in the English language and literature courses that I taught. I assigned reading and writing tasks that addressed, for example, the construction of the highly controversial fourth nuclear power plant, the loss of Taiwan's coastal wetlands to the petrochemical industry, the high concentrations of industrial metals in agricultural soils, the illegal dumping of industrial waste in major tributaries, the proposed dam projects in the south, the bulldozing of farm lands to make way for industrial science parks in central Taiwan, and the ongoing cementing of Taiwan's coastline (approximately 50 percent of which now is lined with cement tetrapods and other industrial-made sea-water barriers).

I had become interested in ecocriticism ten years earlier, when I enrolled in a doctoral degree program in the Department of English at the University of Texas and taught English language and literature courses in the English Department and Rhetoric and Writing Department. In 2005, after I received my doctorate, I returned to Australia and found work teaching ESL (English as a Second Language) and TESOL

(Teaching English to Speakers of Other Languages) courses in several institutions. I taught primary and secondary school teachers from the Republic of Korea under a private TESOL teacher training program, exchange students from India enrolled in the Language Intensive Courses for Overseas Students (ELICOS) program at Northern Melbourne Institute of Technical and Further Education (NMIT), "new arrival" immigrants from countries in Africa and the Middle East enrolled in the Migrant English Program (AMEP) at Victoria University, and "returning" senior citizen students enrolled in the Courses in Spoken and Written English (CSWE) certificate program at Moreland Adult Education (an adult community and education "neighborhood house" located in the northern Melbourne suburb of Coburg). Although the curriculum frameworks did not evidently or primarily reflect the concerns of ecocritics, they were flexible enough so that I could include in the syllabi for the courses assessment tasks that addressed environmental concerns. In doing so, I first had to answer the question of why "ecocritical English" was important to teaching English language and literature.

As Richard Kerridge asks the aforementioned question in "Ecocriticism and the Mission of English," it is whether it is appropriate to *use* English (English Literature, ESL, TESOL, and so forth) to teach environmental awareness and responsibility. As Jonathan Bates had asked the question in an early important ecocritical study of British poetry and prose titled *Romantic Ecology: Wordsworth and the Environmental Tradition*, it was, "do we really need a 'green' literary criticism to go with our lead-free petrol and ozone-friendly deodorant?" (qtd. in Kerridge 14). One answer, as Kerridge supplies, is the magnitude and urgency of the environmental crisis that we face today demands a green literary criticism (14). Another answer is offered by Erin James in "Teaching the Postcolonial/Ecocritical Dialogue." James brings up the question of what kinds of texts should teachers use or rely upon to teach "ecocritical English." Adding to Kerridge's general definition of "ecocritical English," James points out that "ecocritical English" is not only about reading literary texts "for *environment*," or about reading texts that are or appear to be "mimetic or realistic depictions of nature"; it also is about reading texts "for *environmentality*," or about reading texts for evidence of the ways in which texts are always embedded in and are produced and shaped by people's "ecological homes" (66).

When they emphasize that "ecocritical English" vitally contributes to and complements other kinds of ecocritical engagements, ecocritics do not claim that "ecocritical English" will bring about "deep changes" in human "desire and behavior" (Kerridge 14) more so than those other engagements. As Shakespeare scholar and ecocritic Simon C. Estok states in a comment on the work of John Felstiner and Bill McKibben, most ecocritics are not primarily intent upon proving that "poetry or ecocriticism will save the earth" (269). However, they do proceed "on the assumption" that these two areas of human enterprise will "surely help" (269). They are doing this in response to ongoing massive environmental degradation.

Climatologists, ecologists, environmental scientists, marine biologists, and so forth, have been collecting and analyzing the physical and other material evidence for global warming caused by human activities such as the burning of fossil fuels, the industrial farming of animals, and the intensive logging of forests. English language and literature scholars study and learn about those concerns under a wide range of research and teaching projects. One project upon which I embarked was inspired by two undergraduate courses that I have been teaching at Tamkang University since 2009. The courses are a required third-year composition course and an elective fourth-year ecology and ecofilm course. The students were required to ecocritically discuss and write responses to poetry and prose and local news articles. The poetry included poems from Miriel Lenore's *in the garden* (2007) and Les Murray's *Subhuman Redneck Poems* (1996). The news articles mostly were from the *Taipei Times*. Other bilingual (Chinese-English) and English-language news media sources included the *Economist, International Herald Tribune,* and *New York Times*. The students also accessed the bilingual website of Wild at Heart Legal Defense Association. This association was founded by Ron Winkler, a former U.S. citizen who has devoted the last thirty years to defending and engaging in environmental activism in Taiwan. In introducing the students to "ecocritical English," or to the work of engaging with English language texts from environmental perspectives, I focused mainly on two theoretical areas of ecocriticism, postcolonial and biosemiotics ecocriticism.

Postcolonial ecocritics engage in kinds of comparisons that were untenable or objectionable to many scholars as recently as a decade ago. They investigate the philosophical and historical connections between colonized human populations and other kinds of colonized populations. In this work, they are acutely aware of and sensitive to the risks that those connections run of repeating or re-endorsing the gross analogies between humans and animals upon which colonizing powers built their empires. Graham Huggan and Helen Tiffin's important study *Postcolonial Ecocriticism: Literature, Animals, Environment* (2010), is an example of that work. They define postcolonial ecocriticism as the examination of "the colonial/imperial underpinnings of environmental practices" (3). They also identify two terms that are particularly useful for addressing the work of Lenore and Murray—two writers who in different ways are keenly and self-consciously aware of Australia's colonial-settler history. The first term, ecological imperialism, appears in an environmental history study titled *Ecological Imperialism* (1986). In it, the author Alfred Crosby identifies and discusses flora and fauna that were accidently or deliberately introduced to the British colony of Australia (1788–1901) as well as into other British colonies, and flora and fauna that were unwittingly or intentionally brought back to the British Isles from the great southern continent. The second term, "environmental racism," refers to the subject matter of "the connections, in theory and practice, of racism and the environment so that the oppression of one is connected to, and supported by, the oppression of the other" (Curtin qtd. in Huggan and Tiffin 2010, 4).

Scholars who work in the area of biosemiotics are finding, testing, and theorizing a wide range of functions or agencies that scientists previously believed were exclusive to the human species. For example, they are finding that plants have functions analogous to the functions in the human brain of memory, sociality, aesthetic selection, and creativity. They also are finding that locomotion or ambulation, a functional capacity that once was considered to be what essentially distinguished animals from plants, is not exclusive to animals. Also, as Wendy Wheeler argues in *The Whole Creature: Complexity, biosemiotics and the evolution of culture* (2006), work in the area of biosemiotics reflects that "words and discourses," or human language, are extraordinarily powerful but "only one aspect of . . . communication amongst other, unconscious and 'gestural,'" kinds of human communication. Those other kinds of communication are "semiotic," or biologically based, and they are shared across biological species lines (17). They comprise the "biological language of the immune system and its conversations with the nervous system and the brain, and the endocrine system," and they are comparable with and analogous to the sign systems found in plants and animals (142).[2]

Lenore (1927) is an active member of the Association for the Study of Literature, Environment and Culture (ASLEC-ANZ), the association through which I first learned about Lenore and became interested in her poetry. Her publications include *in the garden* (2007), published by the South Australian based publisher Wakefield Press. Its ostensible subject of inspiration is the historic Botanic Gardens in Adelaide, the capital city of South Australia. The second section of *in the garden*, which focuses on the subjects of the trees in the Botanic Gardens, begins with an excerpt from the novel *The Color Purple* by the African-American novelist Alice Walker. Sugar Avery ("Shug") says to her friend Celie, "You ever notice that trees do everything to git at tention that we do, except walk" (qtd. in Lenore 2007, 11). A reader might point out that Shug commits the pathetic fallacy of attributing human traits—namely, affective traits—to nonhuman species when she tells Celie that trees *want* to be noticed ("do everything to git attention"), but a biosemiotics scholar would point out that Shug's statement reflects current research about plant agencies that are analogous to human affective traits. Shug also tells Celie that plants, unlike animals, do not "walk." Here, one might point out that whereas in the past botanists and other plant scientists believed that plants could be distinguished from animals by their inability to "walk," scientists today are challenging that distinction.

A so-called speculative or imaginative writing, *in the garden* ecocritically complements the theoretical work of scholars who specialize in such areas of inquiry as biosemiotics and postcolonial studies. It speaks for plant histories and "experiences" of transportation, colonization, servitude, imprisonment, loneliness, indigenization, naturalization, isolation, hybridization, obloquy, and loss. One of the most haunting poems in the collection that provokes postcolonial and biosemiotic ecocritical readings is "The Oak" (Lenore 2007, 15). The speaker narrates the natural life of a Northeast American species of oak (*Quercus macrocarpa*), brought to Australia and planted in the Botanic Gardens of South Australia at the time of federation when the governor of South Australia was Hallam, Lord Tennyson, the son of Alfred, Lord Tennyson, named after his father's beloved friend Arthur Hallam (15, 81).

Branches of the huge Burr Oak
(*Quercus macrocarpa — East North America*)
touch the ground on three sides
the fourth is cut away
for us to sip tea at white plastic tables

Those plebian birds (or so I think)
black ducks and seagulls pick up crumbs
but the tree is *noble*:

Planted by the State's Governor in 1901
Hallam, Lord Tennyson
who carried always (did his mother agree?)
his father's love for his friend

reminding me that poetry, love and iced coffee
are the centre of my city[3]

"The Oak" expresses in understated and prosaic language, which is characteristic of *in the garden* as a whole, humans' nonchalant uses of and attitudes toward the non-human world of flora in the period of colonization and the mapping and parceling out of green space in the neo-colonial twentieth and twenty-first centuries. One of the Burr Oak's long arms has been chopped off to provide a more comfortable seating space for the humans who frequent the Botanic Gardens. It has been "cut away / for us to sip tea at white plastic tables." The description of the severed branch and the tree's remaining long draping branches evoke, in addition, Australia's identity as one of the most extensive reaches or "branches" of the British Empire and the severing (to a limited extent) of the ties between Australia and the United Kingdom in 1901 under the act of federation. The reference to Australia's identity as a British colony (1788–1901) also is hinted at in the poem's title, "The Oak," which brings to mind the iconic symbol of British identity, the common species of English Oak (*Quercus robar*). Such subtle symbolism is reinforced by the reference in the poem to Alfred, Lord Tennyson, a poet who lived in England in the period when "more than a quarter of all the territory of the surface of the earth" was directly or indirectly claimed by the British Empire (Greenblatt 2012, 1018).

The final two lines of "The Oak" seem to abruptly sweep aside the implicit post-colonial ecocritical arguments advanced in the first ten lines as well as to dismiss the main subject of focus of those arguments, the Burr Oak. The speaker finishes that "the centre" of Adelaide for him or her is not the Botanic Gardens let alone their resident Burr Oak, which has grown too big and become too cumbersome. Rather, "the centre" is "poetry, love, and iced coffee." He or she speaks for a visitor who is as indifferent to the tree as is the administrator who instructed a gardener to lopsidedly trim the tree to make room for plastic chairs. For that visitor and administrator, the Burr Oak and the Botanical Gardens no longer count for much. They are anachronistic, out of place, and in the way of culture and modern life. Yet, the sudden dropping of or turning away from the subjects of the Burr Oak and gardens in the final two lines puts into question the disregard or defilement that they seem to serve.

"The living fossil" is another poem from *in the garden* that speaks to the work of postcolonial and biosemiotic ecocritics (2007, 14). The poem's title references one of the world's oldest tree species, *Wollemi nobilis.*

Once animals were captive
in the centre of the Garden
now a pine tree is behind bars

Found in the Greater Blue Mountains
the precious *Wollemi nobilis* stands
in its cage on the slope above the lake

Known only from fossils till 1994
rarity and phylogeny combine
to make this a remarkable plant

Not named for its character, Gondwana's
forgotten pine honours David Noble
the Park Ranger who found it

Healthy after its transplanting here
it will test, with a Mounty Lofty sibling
The effects of habitat

A child puts his fingers through the bars
The pine does not bite[4]

As the poem relates, a park ranger first discovered the rare species of *Wollemi nobilis* in 1994 in a region of the Blue Mountains in (the state of) New South Wales. This region now is a designated national park, Wollemi National Park. The poem's speaker describes the tree as "Gondwana's forgotten pine" (14), alluding to the late discovery by Anglo-European Australians of the species and to the tree's ancient lineage, which traces back two hundred million years to the time of two supercontinents. One of these, Gondwana, joined the land mass of Australia to other land masses in the southern hemisphere—namely, Africa, Antarctica, Madagascar, and South America ("The Wollemi Pine"; "Gondwana"). As Lenore records, the *Wollemi nobilis* tree that was brought to Adelaide's Botanic Gardens twenty years ago was put "behind bars."[5] This small but telling detail emphasizes two opposing sets of environmental attitudes toward the unknown world of flora and fauna. The first kind associates with curiosity and delight in that world and the effort to preserve it by cultivating and establishing green spaces such as botanical gardens and other parks. The second associates with fear and hatred and contempt of that world, the attempt to sequester and control it, and the neglect of it (as this is implied by an endnote by Lenore) when it is no longer perceived to be valuable. Lenore gives more attention in her poem to the second set of attitudes for the reason, as I see this, that these are less spoken about, learned about, or debated by many people and their elected governments. The focus on those attitudes in "The living fossil" appears in the final lines. A child "puts his fingers through the bars" of the cage that separates the *Wollemi nobilis* from its human spectators: "The pine does not bite" (14).

Lenore's *in the garden* (2007) quietly and subtly speaks to the interests of "ecocritical English" in the context of a historic botanical gardens located in the capital city of the state of South Australia. Les Murray's *Subhuman Redneck Poems* (1996), a very different collection, also carries ecocritical content. Murray is considered by many critics to be Australia's most technically accomplished poet (Kane 79) as well as its most distinguished pastoral poet (McCooey 901) and "bush bard" (Chiasson 136). His poetry stands "at the fore" of "recent Australian literature" in which "the landscape" is "the main protagonist" (Rothwell 33).[6] One of the poems in *Subhuman Redneck Poems* that represents this aspect of Murray's work is "Late Summer Fires." It implicitly argues that "white" Australian identity cannot be separated from "black" Australian identity. In addition, it is about an ubiquitous feature of the Australian landscape, bush fires, and it evokes current debates about bush fires that relate to the problem of deforestation.

In "Late Summer Fires," which ends with three climactic words, "the Aboriginal flag," the description of a bush fire—a brutal and unrelenting force that obliterates pastoral lands ("paddocks"), burns green grass black, and makes kilns of hollow tree stumps and smouldering "logs" of cattle—functions primarily as a metaphor for Australian identity as this cannot be separated from Australian Aboriginal identity, the struggle by Australian Aboriginals for political rights, and the Australian government's belated recognition of Australian Aboriginals. Except for "white," the only color adjectives that appear in the poem (red, yellow, and black) are the same as those found in the Aboriginal flag. Murray also uses those color adjectives to describe the subject of a bush fire. Together, both subjects—flag and fire—form an image of an Australian identity that is deeply shaped by Australian Aboriginal identity, by the political struggle for Australian Aboriginal land rights, and by colonial pastoralist enterprises that fanned the flames of environmental as well as political violence.

Australian Aboriginal history is closely intertwined with the history of the exploitation of the Australian continent by Anglo-European pastoralists. Colonial-settler enterprises transformed the Australian continent into vast stretches of wheat farm and cattle and sheep stations. The oldest environments were not completely beaten ("walloped") by those enterprises—"Even this day rolls over."—but the lack of understanding about and the disregard for "the bush" caused tremendous environmental degradation. Today, this continues to be seen in the periodic relaxation of building zone regulations, the political lobbying of real estate developers to industrially develop or urbanize areas that are prone to bush fires, the aggressive acts of mining companies and nuclear power interests, and the loss of and disrespect for indigenous knowledge about how to work with instead of fight against bush fires.[7]

One can teach poems such as Murray's "Late Summer Fires" and Lenore's "The Oak" and "The living fossil" not only to expose students to a wide range of English literature but also to encourage students to develop and cultivate environmental awareness of and knowledge of lost, threatened, and saved green worlds. In two courses that I regularly teach in the English Department at Tamkang University, a required third-year composition course and an elective fourth-year ecology and ecofilm course, I introduce students to ecocriticism. Recently, I have been focusing on the sub-areas of postcolonial and biosemiotics ecocriticism. I assigned assessment tasks that required the students to discuss and write ecocritical responses to Murray and Lenore's poems and to compare the ecocritical arguments in the poems to the arguments found in news articles that focused on an environmental issue. Most of the news articles were from the *Taipei Times*, an online and print news publication that is remarkable for its generous coverage of local environmental issues, is stocked by most convenience stores, is free online, and is suitable reading for undergraduate students. Students also read and discussed articles published in the *International Herald Tribune* (renamed the *International New York Times* in 2013), the *New York Times*, and the *Economist*.

For one assessment task, the students compared the arguments in the poems to an argument made in the special issue of *The Economist*. Titled "the World's Lungs," the special issue included a report on the world's diminishing tropical rainforests. The students also compared the arguments in the poems to the arguments found in several *Taipei Times* articles and other local news articles about the problem of deforestation in Taiwan. They learned about the role that rainforests and other tree species play in reducing, storing, and recycling carbon dioxide, a "greenhouse gas" in over supply in the present century. Reading and discussing David Momphard's "Protecting Ancient Giants," they learned about efforts by the Atayal Aboriginal people of Smangus to save the Formosan cypress where it survives in what is now Hsinchu County in central Taiwan. The Formosan cypress (*Chamaecyparis obtuse var. formosana*) and red cypress (*Hamaecyparis formosenis Matsum*) are two of the world's six remaining cypress species ("Protecting the Ancient Beauty of Nature"). This work led them to the discovery of a recently released documentary film, *A Year in the Clouds*, which also is about the Atayal's attempts to halt ongoing ecocide of the Formosan cypress, a tree that is sacred to them. The students learned that illegal logging activities are threatening other tree species that were able to survive the period of "hyperdeforestation" during the time of Japanese colonial rule (1895–1945) and after 1949, when the Chinese nationalists took control of Taiwan and continued to log forests at a frenzied rate (Thornber 85).

The students also read, discussed, and wrote responses to several poems by Murray and Lenore by comparing the arguments in those poems with the arguments found in an article published in the *New York Times* as well as in two *Taipei Times* news articles titled "Tropical Mangroves Excellent at Storing Carbon, Study Says" (anonymous author) and "Kenting Park's Coral Reefs Disappearing" (Lee). The news articles were about the world's so-called blue carbon species such as coral reefs, algal reefs, and mangroves. In Taiwan, those species are found in, for example, Pingtung County at the southernmost tip of Taiwan, Taoyuan County at the northern tip of Taiwan, and along the Tamsui River in New Taipei City County. As the author of "Tropical Mangroves Excellent at Storing Carbon" reports, blue carbon species match and even outpace their taller terrestrial green-carbon peers in terms of carbon dioxide consumption and storage and conversion of carbon dioxide to oxygen. As staff reporter I-chia Lee writes about in "Kenting Park's Coral Reefs Disappearing," coral reefs in Taiwan have shrunk dramatically and those in Kenting have disappeared by more than 60 percent in the last twenty years. In "A Coral Reef Endures against the Odds," Jonathan Adams reports on the research of Chang-feng Dai, a coral expert at the Institute of Oceanography, National Taiwan University. According to Dai, Kenting's coral reef coverage was as high as 75 percent to 80 percent in 1988, or almost twice as what it was thirty years later in 2008. He predicts that "fewer than 50 of Kenting's 300 species will survive by 2050, with few or none left alive by 2100" (Adams).

Another article that the students responded to by both writing about and delivering in the form of a presentation, is I-chia Lee's "Activists Decry Water Pollution." It is about the algal reefs in Taoyuan County, located in the north of Taiwan, not far from where the Taoyuan International Airport is located. Algal reefs might appear unprepossessing or hardly worth bothering about in comparison to their bright coral reef peers, but like the coral reefs, they are fantastic storers and recyclers of carbon. The algal reefs of Taiwan represent 10 percent of the world's total extant algal reef population. Those in the north, on the coast of Taiwan's Taoyuan County, home to the island's largest international airport, make up 40 percent of the total algal reef populations in Taiwan. They have dwindled from a length of twenty-seven to five kilometers and are projected to completely disappear in the next two decades as a consequence of decades of under- or poorly regulated industrial development (Lee, "Activists Decry Water Pollution"). In learning about this issue, the students also learned about the mangroves lining the Tamsui River in the suburb of Hongshulin, next door to the suburb of Tamsui. The mangroves, which now are protected, are a short walking distance from the main train station that students use to reach Tamkang University. The students learned that the rates of attrition of mangroves elsewhere in Taiwan are similar to the rates of loss of its forests, which fell from 95 to 55 percent coverage of the island in less than a hundred years.

The Formosa cypress contrasts markedly with, yet also shares, some of the features of the arguably equally incomparable, exquisite, jade-green, bamboo-like, lofty, and human-made (or *anthropogenic*) Taipei 101 skyscraper, until recently the world's tallest building and until today perhaps more well known to many people than the Formosa cypress. It is a magical human-made (anthropogenic) "tree" and could be compared with the astonishing feat of the nonhuman-made (ecogenic) Formosan cypress "sky-scraper." Such comparisons raise debates about whether it is worth preserving the latter or replacing it with such artefacts as the tall, bamboo-like Taipei 101 building. As the ecocritic Timothy Morton would argue, both totemic kinds of trees or skyscrapers, one human-made, the other nonhuman-made (*ecogenic*), are worth venerating; however, he would also probably agree that humans as a species need to consider slowing their frenetic replacing of the former with the latter. Although Morton argues that popular environmental thinking falsely sets up the natural world, so-called "Nature," as a "reified thing in the distance, under the sidewalk, on the other side where the grass is always greener, preferably in the mountains, in the wild" (Morton 3), and that humans have been "terraforming" the earth "all along" and there are very few places on the planet that have *not* been modified by humans in some way, he calls attention to how humans also have been terribly overwriting the face of earth, especially so since 1949 (133).

The so-called natural world does not exist or not nearly to the extent it existed even sixty years ago. Ecological thinking has as much to do with preserving what is commonly or conventionally recognized as the natural world—flora, fauna, trees, mountains, oceans, rivers, and so forth—as it has to do with embracing new forms of "natural"—namely, forms heavily stamped by the human. Above all, ecological thinking emphasizes the value of coexistence of species (Morton 4). In the modern era, particularly in the decades after 1950, human-made and human-generated environments, systems, and things have escalated. We may wish to begin giving more consideration to the artwork, fabrications, machineries, and affective traits of species besides those of our own. "Ecocritical English" is an area of English Studies that reflects the attempt to realize that wish.

When I first taught Murray's "Later Summer Fires" poem in Melbourne, Australia, it was to a class of non-native English speakers in an adult community education facility in a northern suburb of Melbourne, where there is a large population of Australians who are from Mediterranean and Middle Eastern linguistic and cultural backgrounds. The class was mostly comprised of retired Greek, Italian, and Lebanese women who had immigrated to Australia in the 1950s to work in the textiles industry in one or other of the clothing and shoe factories that once made up the backbone of the Australian economy. We had just finished a reading and writing project on the Australian Aboriginal writer Nugi Garimara's (Doris Pilkington's) novel *Follow the Rabbit-Proof Fence* (1996) and on director Phillip Noyce's film adaptation of Garimara's novel (2002). We also had been following the recent news coverage of the Australian government's official apology to Australian Aboriginals and learning about an earlier important political moment, the Australian government's formal recognition of the Australian Aboriginal flag. When I began to teach Murray's poem, I could not do so without thinking of Australia's histories of ecocide as well as genocide. When I introduced Murray and Lenore poetry as well as other Australian poets to my students at Tamkang University, I did so by ecocritically teaching the poetry alongside local news articles about Taiwan's own histories of ecocide and genocide. English studies in Taiwan must not turn a blind eye to or dismiss "ecocritical English, "which brings attention to current pressing environmental problems. Probably, English studies will not be able to do so, for it is increasingly difficult to dismiss, as being secondary concerns of English, the primary concerns of environment.[8]

NOTES

1. For the history of ecocriticism at Tamkang University, see Huang.

2. See, also, Wheeler, "The Biosemiotic Turn: Abduction, or, the Nature of Creative Reason," and Morton, *The Ecological Thought*, 13, 114. In a looser manner of argumentation, Morton points to research about the constitutive role that biology plays in consciousness, memory, desire, aesthetic reasoning, and so forth.

3. Lenore, Miriel. "Living Fossil," *in the garden*. Kent Town, South Australia. Wakefield Press, 2007. Print.

4. Lenore, Miriel. "The Oak," *in the garden*. Kent Town, South Australia. Wakefield Press, 2007. Print.

5. Today, the tree is no longer cordoned off from human visitors (Lenore 2007, 81).

6. For an ecocritical summary of pastoral, see Gifford, *Pastoral*; Gifford, "Pastoral, Anti-pastoral, Post-pastoral." For a postcolonial ecocritical analysis of pastoral, see Huggan and Tiffin, 82–132.

7. The word "subhuman" in the title of the collection in which "Late Summer Fires" appears evokes the history of race relations between Aboriginal and AngloEuropean Australians in the context of a famous lecture delivered by Bernard Smith in 1980. In that lecture, the first of the distinguished Boyer Lectures, Smith used the word "subhuman" in reference to white racist attitudes toward indigenous Australians and to, in particular, an Australian Aboriginal known as Truganini. Murray's "subhuman" distinguishes a white underclass of Australians—namely, poor, rural, and working-class Australians—but the sense in which Smith used it pervades the sense in which Murray uses it. Smith states, "A spectre has haunted Australian culture, the spectre of Truganini. . . . [S]ince 1788 Aborigines have [continued to be] treated in their own county as if they were sub-human. . . . Truganini's story must stand, in these talks, for all those that will never be written" (qtd. in Taylor 26).

8. For English pedagogy and ecocriticism as this concerns the area of secondary school education, see Matthewman, *Teaching Secondary English as if the Planet Matters* (2011).

REFERENCES

Adams, Jonathan. "A Coral Reef Endures against the Odds." *The New York Times*. Web. 6 October 2008.

Bate, Jonathan. *Romantic Ecology: Wordsworth and the Environmental Tradition*. London and New York: Routledge, 1991. 9. Print.

Beyond Beauty: Taiwan from Above. Directed by Chi Po-lin. Produced by Hou Hsiao-hsien. Activator Marketing, 2013. Film.

Chiasson, Dan. "Fire Down Below. The Poetry of Les Murray." *The New Yorker* 11 and 18 (June 2007): 136–39. Print.

Crosby, Alfred W. *Ecological Imperialism: The Biological Expansion of Europe, 900–1900*. Cambridge, UK: Cambridge University Press, 1986. Print.

Curtin, Deane. *Environmental Ethics for a Postcolonial World*. Lanham, Maryland: Rowman & Littlefield, 2005. 145. Print.

Estok, Simon C. "Activist Ecocriticism: An Introduction." *Forum for World Literatures*. 6.2 (June 2014): 261–71. Print.

Felstiner, John. *Can Poetry Save the Earth? A Field Guide to Nature Poems*. New Haven, CT: Yale University Press, 2009. Print.

Garimara, Nugi (Doris Pilkington). *Follow the Rabbit-Proof Fence*. St. Lucia: University of Queensland Press, 1996. Print.

Gifford, Terry. *Pastoral.* London: Routledge, 1999. Print.

———. "Pastoral, Anti-pastoral, Post-pastoral." *The Cambridge Companion to Literature and Environment.* Edited by Louise Westling. London, Cambridge University Press, 2014. 17–30. Print.

"Gondwana. " *Encyclopædia Britannica Online.* Encyclopædia Britannica Inc., 2012. Web. 13 December 2012.

Greenblatt, Stephen ed. *Norton Anthology of English Lit.* Vol. 2. 9th ed. New York: W. W. Norton, 2012. Print.

Huang, Peter. "Ecocriticism in Taiwan." *Ecozon @* 1:1 (April 2010): 185–88. Print.

Huggan, Graham, and Helen Tiffin. *Postcolonial Ecocriticism: Literature, Animals, Environment.* London and New York: Routledge, 2010. Print.

James, Erin. "Teaching the Postcolonial/Ecocritical Dialogue." *Teaching Ecocriticism and Green Cultural Studies.* Edited by Greg Garrard. Houndmills, Basingtoke, Hampshire: Palgrave Macmillan, 2012. 60–71. Print.

Kane, Paul. *Australian Poetry: Romanticism and Negativity.* Cambridge University Press, 1996. Print.

Kerridge, Richard. "Ecocriticism and the Mission of English." *Teaching Ecocriticism and Green Cultural Studies.* Edited by Greg Garrard. Houndmills, Basingtoke, Hampshire: Palgrave Macmillan, 2012. 11–23. Print.

Lee, I-chia. "Activists Decry Water Pollution." *Taipei Times.* Web. 10 October 2012.

———. "Kenting Park's Coral Reefs Disappearing." *Taipei Times.* Web. 12 September 2012.

Lenore, Miriel. *in the garden.* Kent Town, South Australia. Wakefield Press, 2007. Print.

Matthewman, Sasha. *Teaching Secondary English as if the Planet Matters.* New York: Routledge, 2011. Print.

McCooey, David. "Les Murray." *Macquarie Pen Anthology of Australian Literature.* Edited by Nicholas Jose. Crows Nest, NSW: Allen & Unwin, 2009. 900–901. Print.

Momphard, David. "Protecting Ancient Giants," *Taipei Times*, 10 August 2003, 18.

Morton, Timothy. *The Ecological Thought.* Cambridge: Harvard University Press, 2010. Print.

Murray, Les. *Subhuman Redneck Poems.* Sydney: Duffey and Snellgrove, 1996. Farrar, Straus & Giroux, 1998. Print.

Rabbit-Proof Fence. Directed by Phillip Noyce. Produced by Phillip Noyce, Christine Olsen, and John Winter. Miramax Films, 2002. Film.

Rothwell, Nicholas. "The Mirror That Creates." *Meanjin* 72.2 (Winter 2013): 18–36. Print.

Taylor, Rebe. "The National Confessional." *Meanjin Quarterly* 71.3 (Spring 2012): 22–36. Print.

Thornber, Karen Laura. *Ecoambiguity: Environmental Crises and East Asian Literatures.* Michigan: Michigan University Press, 2012. Print.

"Tropical Mangroves Excellent at Storing Carbon, Study Says" *Taipei Times.* Web. 5 April 2011.

Walker, Alice. *The Color Purple.* New York: Pocket Books, 1982. 179. Print.

Wheeler, Wendy. "The Biosemiotic Turn: Abduction, or, the Nature of Creative Reason." *Ecocritical Theory: New European Approaches.* Edited by Axel Goodbody and Kate Rigby. Charlottesville and London: University of Virginia Press, 2011. 270–82. Print.

———. *The Whole Creature: Complexity, Biosemiotics and the Evolution of Culture.* London: Lawrence & Wishart, 2006. Print.

"The Wollemi Pine: A Very Rare Discovery." The Royal Botanic Gardens and Domain Trust. Web. 2 February 2013.

"The World's Lungs: Saving the Forests." *The Economist.* Web. 23 September 2010.

A Year in the Clouds. Directed by Dean Johnson and Frank Smith. Produced by Taiwan Public Television Service. Infocus Asia, 2011. Print.

6

Interacting with "Other" Places in Modern Poetry through Hypermedia

Emerald C. L. Ku and Patricia Haseltine

Modern American poetry is usually not introduced to students learning English as a foreign language in Taiwan. The poetry of William Wordsworth and Emily Dickinson is more familiar to our students than the poetry of Allen Ginsberg, Charles Olson, James Merrill, Elizabeth Bishop, and Robert Lowell. However, the poetic images these poets create, picturing foreign places in a contemporary mindset, may offer students in Asian countries valuable cultural insights. By reading poetry about place in a hypertext[1] environment, students may enter the poets' meaningful interpretations of land- and townscape to obtain a virtual experience of the places imaged. In this study, we concentrate upon two poets from New England, Elizabeth Bishop and Robert Lowell. Lowell's poem "Skunk Hour," with a Maine village locale, is a direct response to Bishop's "The Armadillo," which records an epiphany on St. John's Day in Brazil. With their acute and anxious visions, both poets penetrate the places of their own being in the world with memorable localized sensibilities.

Thus focusing on the relations of place and poet, this study postulates what happens to implied hypertext readers in Taiwan as they navigate the spaces of such poems and their intertexts. That is, we are more interested in what the foreign language students can learn about the environment of the poems and how they can approach the intimate connections the poets have to places than in what we are teaching in terms of the critical interpretations of these works. We suggest ways the alien places can become meaningful in the experience of students located in Taiwan as they navigate through the poems on a virtual journey to the "other" places depicted and beyond since such voyages necessarily decenter the primary texts themselves. The experience of each student is a unique voyage because each will search in hypertext choices of her own in interacting with the poems and build up a personal store of instances for encountering the cultural matter informing each poem.

In the hypertext environment, the shared space of these two poems is rooted in the history of the poets. Robert Lowell is Bishop's poet friend as well as a colleague for whom she herself felt close affinity (Lombardi 70). Yet, as Goldensohn has perceptively indicated, the relationship is a complex and rich narrative needing to be told on several levels: "each in the fashion of his or her nature supported the other's career, each was influenced both directly and indirectly by the other's poetry, and each held the other in some inner and inviolable chamber of the mind as a figure for contemplation" (163). Indeed, after Bishop met Lowell in Boston in 1957, they established a life-long friendship in very subtle ways found in the publication of their letters in 2008 (Bishop). Most significantly, as poet friends, they supported and influenced one another's poetic creativity.

The story behind and relationship between the two poems, Bishop's "The Armadillo" and Lowell's "Skunk Hour," in a sense is an epitome of the poets' relationship. Firstly, Lowell's poem is based on an event Bishop happened to be involved in. Bishop once confided to her students that "Skunk Hour" was about a scene she shared with Lowell one summer when she was visiting him in Castine, Maine. As she described it, "we heard some noises out in the back yard and when we looked we saw a family of skunks going through the garbage" (Gioia 150). Moreover, there is a more subtle relationship between these two poems. Though David Kalstone points out that there are many differences between the poems—one is more confessional and historical, the other is purely a whole natural scene (187), Lowell himself clearly confessed his indebtedness to Bishop for her poetic influence on him. After Bishop dedicated "The Armadillo" to him, Lowell wrote "Skunk Hour," which bears the dedication "For Elizabeth Bishop," and claimed that it was partly modeled on her poem "Armadillo":

> When I began writing "Skunk Hour," I felt that most of what I knew about writing was a hindrance. The dedication is to Elizabeth Bishop, because rereading her suggested a way of breaking through the shell of my old manner. Her rhythms, idiom, images and stanza structure seemed to belong to a later century. "Skunk Hour" is modeled on Miss Bishop's "Armadillo." (Goldensohn 184)

The influence seems to extend more widely and deeply as we expand upon the intertextuality. Lowell "began to feel that real poetry came, not from fierce confessions, but from something almost meaningless but imagined" (186). Here the poet seems to suggest that when meaningfulness is gone, we look out on and accept frightening reality, which can be imagined. That is perhaps why Lowell in "Skunk Hour" starts with drifting description but ends with a close-up of an animal epiphany, just as Bishop does in "Armadillo."

Yet, in the unending and fruitful dialogic relationship the indebtedness is mutual; that is, Bishop was also inspired by Lowell to some extent as some critics suggest. For example, Goldensohn notices that the italicized endings in "Armadillo" "signal Bishop's attempt to bring her book into unity, and probably to emulate the blaze and finish that excited her in Lowell's *Life Studies*." It is here that "her solutions turn on uses of narrative. Her immersion in prose has given her rolling, incremental masses of material, sighting in a range of places for her from Nova Scotia to Brazil, and augmenting detail without sacrificing plot momentum" (190). Indeed, the imagined and localized narratives of the two poets have a similar plot momentum and collection of images that invite comparison in a hypertext-oriented approach. For our readers then each of these poems is not only a companion piece found in the environment of the other poem, but has a dialogical relationship that can be established as well through hypertext.

In "doing English" by teaching English poems about place in an Asian country, we agree with Tuan Yi-Fu and Lawrence Buell on the importance of place in the human consciousness. Tuan, a cultural geographer, explains how place is significant to human beings. In *Space and Place,* he astutely analyzes how humans and animals relate to places through our various senses: "Human beings not only discern geometric patterns in nature and create abstract spaces in the mind, they also try to embody their feelings, images, and thoughts in tangible material." Not only is space defined by objects and places, "objects and places are centers of value. They attract or repel in finely shaded degrees. To attend to them even momentarily is to acknowledge their reality and value" (17–18). Tuan also introduces the useful term "topophilia" (4), for the feelings human beings have for place, in his book titled *Topophilia.* The Maine resort town that is in a state of decline largely repels the speaker in "Skunk Hour," who tends to blame himself, not his changed environment, for this feeling of malaise. The natural Brazilian landscape with its lanterns floating up into the night sky is felt to be a place of attraction to the speaker of "Armadillo" until she registers the pain and anguish of small animal creatures hurt by falling fire lanterns on this religious feast day of St. John. Often the objects and places in poems such as these may "lack the weight of reality" (Tuan 18) for students of literature who can only read about them at a distance. We cannot hope to lessen that real distance; however, our students may experience a virtual journey that replicates in the hypertext space the poets' experience of these objects and places

Concentrating on the New England heritage stemming from the transcendentalism of Ralph Waldo Emerson and Henry David Thoreau, Buell analyzes the discourse on the human environment as it occurs in literature. Emerson's *Nature* was, according to Buell, one of the first works of ecocriticism. Thoreau's complaints about the people of the village of Concord resonate with Lowell's feelings about his own village and provide a useful comparison of solitary life in the New England town as opposed to solitary life in nature.

In *The Future of Environmental Criticism,* Buell distinguishes several spatiotemporal coordinates or, to use Tuan's term, kinds of "topophilia." Drawing on phenomenology, first, Buell claims that one of the most significant experiences of place is that of being in the traditional community where we live, from which there are concentric circles of diminishing emotional attachment. So, one could say, if a poet introduces a place of community familiar to her experience in New England, it will be quite remote from our own students' experience and their own emotional attachment to places where they grew up in Taiwan. The second of Buell's attachments to place is that of the archipelago or "dispersed islands of activity." These are places we go to frequently away from our home community for work or vacationing or visiting. Although Brazil may be farther afield than Buell's categorization allows, Bishop lived as an expatriate there for fifteen years. For most of the students who have not traveled so far, Brazil and New England are equally remote. For them, the poems are literary places, which belong to Buell's third category of "virtual places" we frequently go to. Long-time students of literature have their favorites, like Odysseus' Ithaca (Homer), Hamlet's castle (Shakespeare), Wordsworth's Lake District, or Joyce's Dublin. For the relation of time with place, Buell posits a category of "place in process" or "*eventmental*" experience of place. We notice the changes in the places we experience at various times in our lives (72–76). And in "Skunk Hour," Lowell registers just such an *eventmental* experience in his sensibilities for a place he returned to throughout his life.

Finally, in *New England Literary Culture: Revolution through Renaissance* in a chapter titled "The Village as Icon," Buell notes how the New England village becomes an icon for writers who have "juxtaposed images of the real and ideal." However, "the effect was not . . . always the critique of the former in the name of the latter (313). Adopting Buell's iconography of the village, one can see Lowell's poem as reinscribing a dichotomy embedded in the tradition of New England regionalism. This is found in Thoreau, who opposes the Concord village by placing a new "center" for himself in the wilderness at Walden Pond. In "Skunk Hour" Lowell moves the New England–self back to a village center where the skunks now have the run of Main Street, while the townspeople of the past are under the ground in their graves and the present townspeople are above ground in their "love-cars." So what Buell calls the "comic grotesque" in the tradition of village writing in New England (chapter 15) is found modernized in Lowell's poem.

Also apropos is Buell's insight that sees such regionalism as presenting "pathographies" rather than "anatomies" since the speaker of "Skunk Hour" says his "mind" is "not right." The challenge we encounter by using the Internet, then, is to present Lowell's poem, not as a tourist attraction, but more as a "pathography," replacing the attractiveness of a Maine resort town with an ironic sense of dis-ease. This, too, is a deeper cultural knowledge of local place for students to understand. The montage-like juxtaposition of graveyard with love cars, the image of run-down mansions and the fairy decorator, who is ready to compromise and marry, upholds this irony of uncanny discomfort in familiar surroundings. A similar challenge is met with in Bishop's "The Armadillo," where the fire lanterns of a religious festival in Brazil disrupt the lives of animals in nature. Fascinated by the touristic picture of lanterns and comparing them to the sky lanterns of Taiwan, students may miss the irony in the juxtaposition of religion and nature, another pathography of place, as they follow hypertext links to discover cultures.

The "pathologies" of poems are often overlooked by foreign students learning new vocabulary and struggling with syntactic variations, activities that ignore the holistic level of contrasts where ironies usually appear. If hypermedia sources are added to the learning task, understanding the poet's self-criticism may still not be communicated. This is because the sources of many images on the Internet promote a positive view of the represented objects the poem is relegating to the background or even criticizing. This Pollyanna view is nurtured by consumerism and capitalism. Therefore, the teacher must still play the role of one who points to the poets' contrasts of images in order to stimulate the sensitivity to ironic perceptions with deeper cultural import. This problem of cultural literacy is a serious one we think must be considered as students acquire knowledge of the language and culture in any medium. Just knowing the vocabulary or Internet lexis of a poem, or even examining the topography of a place, is not enough to have language sensibility and cultural literacy. The advantages and limitations of the hypertext voyage have been thoroughly researched in the last two decades, but major questions about reception and use in foreign literature classes still need to be scrutinized as the technology develops.

HYPERTEXT EXPERIENCE: INTERTEXTUAL SPACE AND INTERACTIVITY

As fellow explorers with our students and, to some extent, designers of hypertext choices for the poems of Bishop and Lowell we have chosen, we postulate an implied[2] hypertext reader who is not a native-speaker of English. Based on the studies of hypertext, we realize that such a reader is different from a more conventional reader of poetry in a book. Herbert Thomas holds that hypertext reading requires a new definition of literacy. He has explained that whereas a traditional text becomes "part of the fabric which constitutes textual situatedness within a wider textual, social, cultural and historical context," a hypertext environment makes "intra- and inter-textual links . . . equivalent." That is, "instead of being confronted by a solitary, discrete text, the user is confronted by a web of interrelated texts. Such intertextuality is a major implication of hypertext." Further Thomas notes, following Spiro et al., that cognitive flexibility is required. The cognitive skills of hypertext users "are characterized by relational, globalizing and holistic thinking" (485). The limits of a book are the same as the limits of a text that is merely read on the computer from left to right, top to bottom. Today many documents and literary works are merely stored on the Internet with only the Internet page copied from the sources. Schreibman outlines the history of hypertext development in literature study, explaining how a major advancement was made when different textual versions could be paralleled. Providing annotations through hypertext links is another way computers enhance the information level of a literary text. Since poetic text is recursive, however, a hierarchical model of encoding such a text is not as important as finding "the principles which govern the reconfiguration of objects" in the computer environment (287).

Following from Bolter's insight that hypertext is to be conceived as a "spatial" field rather than as a "book" in which linearity is commanded, Dwight and Garrison warn, "the hyperworld we imagine has no fixed centers" (713). The computer is just one site in which such decentering of the subject occurs, but it is the most disruptive and therefore has the most potential for empowering the reader with choices. In a conversation with Richard Andrews using the metaphor of the city of text, Wendy Morgan points out that "if one's talking in spatial terms, [hypertext] doesn't allow for a central or top-level text with underpinning materials or stuff that in a print text would be relegated to footnotes at the back." Whatever is "on the screen is in a central place. It puts in question what is text and what is paratext" (89–90). George Landow likens hypertext to a collage made of different kinds of materials and to a "changeable montage" (195–197). Silvio Gaggi (101–110) attempts to explain the decentering of a literary work within the critical framework of post-structuralism.

When reading an engrossing novel, the reader's immersion is a common phenomenon. In this type of virtual experience what is read in the linear text guides the reader's responses. Ryan has pointed out, there is a distinction between "immersion" and "interactivity." Hypertext intertextuality even "blurs the boundaries between text and reader" (Landow 4) as the reader becomes a more active creator of what she reads.

The influence of the designer of the links for hypertext usage restricts the reader, too. Bromme and Stahl confirm the spatial metaphor for hypertext reading and view the frequently used term "navigation" as an apt one for making spatial associations (117). When students themselves construct the body of hypertexts, their activity "can contribute to a comprehension of the concepts and of conceptual differences within the subject matter. Moreover, "thinking about . . . necessary links requires a processing of semantic relations between the concepts explained in single nodes. A thoughtful application of links can thus contribute to the comprehension of semantic relations." Finally, "an author has to comprehend the content structure of the subject matter. Because of their multilinearity, hypertexts can be read in different ways, so the hypertext author has to anticipate possible audience perspectives in order to create flexible ways of reading the hypertexts (116). Giving students the role of designing the hypertext themselves would stimulate more critical thinking.

In a design for a learning program on the Internet, there should be a dimension of interactivity. Interactivity with the text can take a number of forms, depending upon the materials made accessible in the design. As designers of hypertext around literary works, Morgan says "our intention is that it will become a read-write hypertext, so that other people's stories can be increasingly brought into it." She further explains the purpose is "to draw other people into a conversation in ways that hadn't been possible except as marginal notes in print books" (90–91). For students and teachers alike, the basic advantage of the Internet as compared with the book is that "in a virtual world experienced as three-dimensional the user knows that reality is not limited to what can be seen from a given position: the outside conceals the inside, the front conceals the back, and small objects in the foreground conceal large objects in the background" (Ryan). Travis moreover explains that hypertext incorporates virtual-reality technology to make the reader become "a role-player in 'real-time' dramatic performance with other readers."

The navigation of implied hypertext readers through the hypertext links to the poems of Lowell and Bishop will depend upon other factors, such as language and internet proficiency. Wen, Tsai, Lin, and Chuang affirm that Chinese students can become more constructive in their learning through the use of hypertext. There is wide difference of opinion, however, about how hypertext use is related to proficiency. On the one hand, McGrath points out a significant pattern: "high-skill subjects took more time, viewed fewer screens, and had the fewest misconceptions of all groups" but "low-skill hypertext subjects took less time, viewed more screens, solved the least number of problems of any group, had the most misconceptions, and made the most nonsequential choices" (515). On the other hand, Balcytiene found that the hypertext format is especially effective for those who have higher capacity to use metacognitive skills to construct a mental model. Metacognition is especially important for adult learners because their prior knowledge is much richer than that of young learners. Molly Travis envisions readers who are sophisticated in cybernetics who can reconstruct texts alone or in performance with others. For them hypertext is an art form that has the potential to offer "the pleasures of virtual immediacy, spontaneity, intricate movement, a rich web of texts in various media (graphic, audio and film) and interactivity . . . in the form of creative agency to reconstruct" (2).

Besides proficiency, another factor influencing hypertext adaptability is the learning style of the individual. Crosby and Iding show learning styles to be related to cognitive performance on multimedia tutors. Multiple Intelligences Theory of Gardner would also be related to the types of background choices made to construct a particular experience of a text. A learner who was higher in linguistic intelligence would make different choices from those who are higher in spatial intelligence, just to give one example (Haseltine and Chen). The research of M. Graff focuses on browsing strategies, distinguishing between the cognitive styles of verbalizers and imagers, finding that there are significant differences in browsing strategies between these two groups: verbalizers use relational strategies while imagers use hierarchical ones. Our implied hypertext poetry readers may well also belong to one of the profiles identified by other researchers as outlined in Lawless et al.: "knowledge seekers," who "pursue information related to content and navigate to screens with material that will increase their comprehension in a systematic way; "feature explorers" who interact more with graphics and sound effects and are interested in the different kinds of screens available; and "apathetic users" who seem to get lost or are making random choices and appear to be "unmotivated to learn from the hypertext" (275–76).

Foreign language learning requires learning new vocabulary, and hypertext provides a quick and easy way of finding out the meanings and pronunciation of words. Tripp and Roby have carried out experiments using a hypertext dictionary and memory. It was found that there are positive effects of using various complementary ways of introducing vocabulary. They also note that experts tend to use the hypertext vocabulary links less than new learners. Thus the teacher in Taiwan as hypertext designer must consider the students' learning needs, such as the need for vocabulary, as well as for background cultural information of various kinds, but also take into consideration that hypertext, as Landow insists, is an open text. In sum, the hypertext reader can be, if he or she wants, a post-structural reader who offers his own choices and constructs his or her own "learning text." To what extent learners are comfortable with searching or need to make background searches determines the degree of interactivity, which in turn will influence the amount or kind of learning obtained from the journey through the virtual places of the poems' hypertext links.

In the following sections we discuss the background to the poems and the range of hypertextual environments students may use for learning about places in English. In both of the poems we have chosen, we identify the images in the poems that help to define the field of the hypertext interactivity available to the student, what are called the sites or "nodes" that may or may not be accessed by individual students. Then we explain how the poets' place-experiences in the more open hypertext environment become a palimpsest of cross-cultural place experiences for the student located at home on a computer in Taiwan.

THE SENSE OF PLACE(S) IN ELIZABETH BISHOP'S "THE ARMADILLO"

Bishop wrote to Anne Stevenson in March 1963 claiming that she was descended "from the Transcendentalists" (Lombardi 70). Moreover, in her prose poem "Strayed Crab," Bishop explains the function of the speaker in her poems: "I believe in the oblique, the indirect approach and I keep my feelings to myself." It is Bishop's metaphorical, understated, and oblique method of feeling-reference that most challenges her readers, both in reading and teaching. Yet, significantly, Bishop is a poet writing to places, as in her work "Letter to N.Y.," and most of the titles of her poems and books show being in particular places has influenced much of her poetry: "Florida," "First Death in Nova Scotia," "Cape Breton," "Paris, 7 A.M.," "Brazil, January 1, 1502," "Arrival at Santos," *North and South, Geography III*. It seems that she is a poet with the power and magic to transfer the geography of the earth into that of the human heart. And it is her style of translating feeling into topography that makes some of her poems good experiments that address the process of accessing the environment represented in the text—the places familiar to the poet who is attempting to translate feeling by familiarizing and making memorable to readers the environment experienced.

"The Armadillo" appears in Bishop's poem collection *Questions of Travel*, in the part called "Brazil," this country being the textual environment represented in the poem. The speaker describes her experience of watching the rising and falling of "illegal fire balloons" during the local festival of St. John's Day in June when Brazilians let balloons or lanterns fly up in the sky as a worship to the saint.

The poem begins by describing a festival experience focusing on the fire balloon ceremony in which fire balloons as "paper chambers flush and fill with light" are sent up into the sky. These man-made lanterns are as beautiful and colorful as the productions of heaven, in which are seen "the tinted" stars and planets—"Venus" and

"Mars." Yet, these man-sent "frail, illegal" paper chambers prove to be "dangerous" when they hit the wind, especially the wrong wind—"the downdraft from a peak." The second part of the poem chronicles that danger after a fire balloon falls against a cliff and destroys the home of a pair of ancient owls, then burns down to disturb the lives of a "short-eared" baby rabbit whose eyes become "ignited," and a "rose-flecked" armadillo, who attempts to escape with head and tail down. The last part of the poem is in the form of an italicized quatrain containing metaphysical images with existential significance. The exclamatory comment, "*Too pretty, dreamlike mimicry!*" initiates the epiphany. What follows is the concrete sensual pathos of a description of the "falling fire" and "piercing" cries of the animals in panic. But the final picture of the armadillo—"*a weak mailed fist / clenched ignorant against the sky!*" (lines 39–40)—brings the religious critique of the human observer and her commiseration with the animal into a transcendental conjunction,

The interpretations of this set of imagery and the possible destruction of the animals or their various habitats vary from critic to critic. Accordingly, while Goldensohn sees that the poet keeps her most pained and bewildered selves under cover in disguise as the armadillo (2), Costello explores the poem's significant use of the grotesque tradition in ways that challenge our conventional responses to life, recalling Buell's observation on the history of New England literature of the village. She focuses on Bishop's juxtaposition of the beautiful with the dangerous, the vital with the morbid, spiritual transcendence with moral indignation. In a sense, the poem dramatically shifts sympathy from the desire for aesthetic transcendence in the fire balloons to the will to survive and the rage against oppression (Lombardi 128, 149).

The geographical perspective on place in Bishop comes by way of the words and images in the first and second parts. It precedes the ambivalent beholder's participation in the repressive forces, first horrified by, and then concerned for the oppressed creatures. By making hypertext choices, students increase their understanding culturally and linguistically and are better prepared for an understanding of the metaphysical ending. In the frightening and beautiful world created in Bishop's poem, students as readers virtualize the tangible real through the words and images they explore in the intertextual environment created by the hypertext.

Interacting with the Sense of Place of the Text of "The Armadillo"

Poetry appeals to sound and sense; therefore, assisting the students with hypertext searches, we teachers start by helping them to access, on YouTube, the reading by Bishop of "Armadillo." Besides helping them attune to the sound of the words, the performance prepares them for some words and imagery they will later explore on the sites to create their own hypertext collection. Moreover, the reading will stimulate their aural sense subjectively in response to the tone of the poem through the voice of the poet as the speaker.

Then, to help students interact with the physical sense of place of the poem, we ask them to surf available map sites on the Internet to virtualize Brazil as a tangible place. First, the map of Brazil searched from Google Maps and other sites, such as Bing Maps, will contribute to their recognition of the locale, even down to detailed street maps. By browsing and searching street maps of the cities in Brazil, students build up a strong visual sense of those places with concrete images, such as metro stations, stadiums, and public-created points of interest. Besides, on Flickr, students can collect photos of church architecture, urban landmarks, storm scenes, rare bird species, beaches, and rural landscapes to add more to their topographical sense of this region in South America.

The annual festivals or carnivals in Brazil are also aspects our students won't miss when doing Internet navigation connections made to showcase Brazil. In these carnivals, aside from colorful costumes, tasteful wine, and fabulous dances, many homemade small hot air balloons are made with papers of several colors, propelled by torches, and launched into the sky during traditional carnivals to celebrate the saints. These festivals have been a staple of Brazilian culture for centuries. In the poem, the fire balloons launched for the event of the Feast of St. John, also known as *Festa Junina*, are of thematic importance. The poem is actually based on the experience of the poet watching the festival of St. John's Day in Brazil.

Moreover, the beautiful sky with Mars and Venus as a background for the beautiful floating fire lanterns is also a point to foreground. When exploring the four stars that comprise the constellation Crux, known as the Southern Cross, via hyperlinks, our students will be provided with numerous fantastic pictures and photos of the starry scenery. In particular, being one of the countries in the Southern Hemisphere, the Southern Cross constellation is incredibly beautiful because of the geographical advantage of Brazil.

While all these pictures and photographs as images create Brazil visually as a virtual beautiful place with various tourist attractions, the designers of hypertext searches for students need to foreground the place consciousness of the poet as a way to help students interact with Brazil as an "other" place for culture teaching.

As predicted by the title of the poem, the armadillo as an animal is a creature from nature meant to represent, along with other animals, its bioregion or habitat. Thus, all the animals mentioned in the poem contribute to a compound synecdoche for Brazil in the place consciousness of the poet. So in the poem, the reader perceives an emotionally charged contrast between the joyous image of the lanterns hopefully rising in the sky and the lit-up eyes of a baby rabbit with a paw so-soft, like a "handful of intangible ash," and the armadillo whose armored shape resembles a "mailed fist clenched against the sky" as it crawls away, leaving the scene of destruction in the firelight.

Indeed, as most critics point out, in the poem "Armadillo," the beauty and the danger are juxtaposed. The fire balloons are described as something dangerous but beautiful, and the animals are rendered with physical simplicity as well as uniqueness, but hurt and terrified as the fire balloons fall flaming to earth. For one thing, the three kinds of animals, cliff-perched owls, short-eared baby rabbit, and rose-flecked armadillo, are depicted in the poem as being wreaked by the shattered fire balloons; thus, they are misfits described with carnivalesque images of panic and misery highlighted by words such as "ignited," "intangible," and "shrieked." Depicting them in a dramatic scene, forced from either their natural nest on cliff homes, or out of a comfortable thicket and easy ground habitat, the poet intentionally has her readers caught up in the terror of the animals.

Therefore, besides the tourist beauty of the city, the dangerous part, hidden beneath some cultural images, also needs to be explored for cultural learning and understanding. That is, based on a firmly defined locale with various beauty, we teachers can introduce some key cultural events as goals for students to do hypertext searches. In the case of the poem "Armadillo," for example, the shattering fire balloons launched on the Feast of St. John's Day to celebrate the Saint can be explored with their threat to natural life as false tributes. Accordingly, the fire balloons in Bishop's poem as a cultural image is a good starting point to help students interact with Brazil through hypermedia. It is particularly significant to do the interacting intertextually since sky lantern festivals are also important cultural events in Taiwan.

Through Wikipedia, the free encyclopedia, it is easy for students to find out that launching fire balloons upward into the sky on the Eve of June 23 and the 24, the day of the Feast of St. John, is a tradition from Roman Catholicism. And through hyperlinks focusing on the sites in Brazil and fire balloons, students will find photos of the typical Brazilian feast, with rich photos, pictures, and written explanations to integrate physical and cultural understanding of the event and its fire balloons. On St. John's Day, people celebrate rural life with featured typical clothing, food, and dance, and at the same time launch the man-made colorful paper fire balloons into the sky as tributes to give thanks to Saint John for bringing the rain to arid Brazil.

Similarly, lantern carnivals are held all over Taiwan; the "sky lantern" festivals, taking place in Pingxi, in northern Taiwan, are the most visually exciting. Most students in Taiwan have the experience of seeing the sensationally beautiful and unforgettable sight of thousands of four-foot-high lanterns soaring upward into the sky all at once. Sky lanterns were invented during the Age of Three Kingdoms in ancient China and passed on from generation to generation as an ongoing tradition. The lanterns are made of paper, and the hot air is produced by an oil lamp suspended beneath the lantern sending it flying in the sky. As a whole, it is like a miniature of the fire balloons presented in the Brazilian *Festa Junina*. Without foregrounding the cultural differences between the two public occasions, students may easily just identify the two events, missing the major thematic significance of the poem, that is, the contrast between the beauty and the danger to natural creatures.

So, first students need to feel the major difference between the two cultures by understanding the spirit behind the events: people in Pingxi releasing the lantern write down their wishes or prayers for peace upon them, while people in Brazil send up the fire balloons in celebratory veneration of the Saint and in a spirit of worship. Then, by doing hyperlinks to the Facebook concerning *Festa Junina*, the history of the feast helps a lot for understanding the flower of Saint John, its music, its decoration, its packages of sweets, its origin, tradition, and spirit. Most significant is that the poem's spirit underscores sympathy, attuning the concern for the disrupted and destroyed natural life connoted in the final quatrain.

Indeed, as the critic Penelope Laurens indicates, though there is no sense of moralizing at all in the poem, the moral dimension is inescapably present through the final quatrain. With doing hypertext searches on the Internet, students can appeal to hypermedia, including reading, listening, and watching videos, to interact with Brazil as the "other" place in the poem "Armadillo." As a result, the hypertextual montage created through hyperlinks, to use Landow's apt description of how this technology works, helps evoke a combination of joy and fear for its readers. Accordingly, such hypertextual juxtaposition in the metaphorical images brings for our students a Brazil that is a beautiful international tourist place on the one hand, and on the other, an "other" place attuned to the consciousness of the poet with the sense of helplessness and existential anguish that are inescapably presented in the poem.

THE SENSE OF PLACE IN ROBERT LOWELL'S "SKUNK HOUR"

The textual environment represented in Robert Lowell's poem is New England. The poet depicts a place on the sea-coast of Maine that is frequented in the summer months by wealthy visitors who like to go sail boating. Real estate, what Sandra Gilbert calls the "ill nature of urban-surburban mid-century America" (73), is the detailed focus of the poem: a local dowager is buying up the old houses that are now deserted. In "Skunk Hour," Lowell elucidates precisely Buell's "eventmental" experience of place: the place is not as attractive to visitors as it once was since the houses are vacant and one old lady buys them up so that no one else will move in. Since she doesn't repair them for resale, either, they just run-down further and become eyesores. The people of the place have accommodated themselves to the state of affairs so much that a gay decorator, whose occupation depends upon the fate of the real estate, has little business and may as well "get married," rather than spend all his energy playing the role of a fashionable "fairy decorator" for the region. Since humans have now become scarce, it is the skunks who now experience the place most fully. They walk down Main Street in broad daylight. They eat from the garbage pails as the sole presence of the quite lonely onlooker-poet observes them. Just as Bishop makes the armadillo and what it ignorantly suffers carry the transcendence of the whole scene, the skunk is a metaphoric synecdoche for all that Lowell feels about the place he depicts.

Criticism of "Skunk Hour" places this poem at a crisis period in Lowell's life experience. Lowell himself in "On Skunk Hour" comments that he thought of it as a Dark Night of the Soul, referring to the great Spanish Renaissance poem of San Juan de la Cruz (131–132). Lowell says he wrote the poem backward starting with the memory of the final mother skunk image (133). Fellow poet John Berryman, for whom Lowell's poem is one of his "subtle, strong, terrible poems," has commented upon the existential anguish in the poem, placing the speaker in a near-suicidal mad state that is counterpoised by a recognition of the skunks foraging in the garbage as a parable, a terrible animal epiphany of the way for continuance and survival.

Mark Rudman, who sees the poet-self at the end of *Life Studies* turning inward and blaming himself for the negative view of the world with the recognition that "I myself am hell," interprets the mother skunk image as having "power without meanness." That is "to accept the skunks is to accept himself" (83–93). William Martz agrees that the skunk image is a hopeful symbol of self-redemption in self-recognition: Man can live with his hell, or gain strength by feeding on the garbage of our civilization" (11–12). Burton Raffel makes an even stronger case for the skunks. They are the speaker's "salvation." The mother skunk is "courageous" and her "physical beauty is compared to the ostrich, a much bigger creature" when she puts her head into the garbage. The savage attack on capitalist society supported by the images of the dowager, the summer millionaire, and the decorator, and the love-cars on the hill by the graveyard is matched at the end by "renewed hope" on the back steps of the home (56–59).

Praise for the final epiphany of the skunk mother image in the poem is not unanimous and underscores the diversity of critical interpretations of the ending. The Freudian explication of the poem by Kramer remarks on the psychological displacement Lowell practices in this important poem that comes at the end of *Life Studies.* By making biographical identifications with Lowell's mother as the dowager, his father as the summer millionaire, and the poet himself as the fairy decorator, Kramer sees the final image of the skunk mother with her children merely as a "therapeutic inspiration." This image is not "an epiphany which leads toward affirmation, but a creative evasion, not an act of self-possession that proves an anodyne, but a creative dispossession" (89–92). Sandra Gilbert's strikingly intertextual interpretation brings out a hidden misogyny in the negativity of the images of the poem's dowager and the mother skunk. Gilbert not only points out the allusion to Marlowe's *Faust* in the Mephistopheles quote, as a number of critics do, but draws parallels of the dowager with Circe, the summer millionaire with the absent king of Eliot's *Wasteland* and Shakespeare's *Hamlet.* All of these figures from other texts could be linked in a hypertextual environment for the poem, but in all of this we as designers must also consider the abilities of our students. They would be much better at feeling what Gilbert describes as the malaise of the "hour" of night in this mysterious poem (70–79).

Finally, critics tend to remark also on the "rich" air at the end of the poem, reminding us that the sense of smell is one multimedia effect for which we cannot yet find a means in the hypertext environment to transmit: the smell of the foxes on Blue Hill or the stink of the skunk and the garbage. The pure imagination of the students must suffice. Gilbert says the smell of the poem itself is "rich" with Modernist and Postmodernist anxiety (79), all of which is quite an abstract conglomeration of literary concepts not so easy to pin down for our students to a virtually tangible environment. In their hypertextual journey, students need not be informed of these psychological insights and intertextual comparisons with other poems in order to perceive discordant notes in the language and imagery used to depict the New

England community circle as alienating and decaying. We surmise that, depending upon their maturity, students, however, may indeed be affected by this imagery and its whole intertextual environment.

Interacting with the Places of "Skunk Hour"

First of all, students can access a reading by Lowell of "Skunk Hour" on YouTube. The performance will attune them to the pronunciation of the words and the tone of the speaker's voice, as well as the audience's audible laughter at the ironic humor in the poem. Linking from the images, students can then explore hypermedia points of interest. Maps of seaside Maine villages, pictures of Nautilus Island where the old lady in the poem lives facing Blue Hill, and other seaside villages around Castine Bay, are a good starting point. Google maps and sites with street views allow students to design their own Internet tours of villages along the coast of Maine. Once the locale itself is firmly defined, images associated with culture can be a goal of the student's hypertext searches. These include at least several different dimensions of how people live in that place: the Main Street, housing and church architecture, the people's livelihood of fishing or decorating and their sports of hunting and sailing, and local folklore and customs.

Although the image of a small town Main Street may seem to be a place concept easily understood cross-culturally, there is a reason for including it in Internet searches for photographs of New England Main Streets, such as those found on Flickr. In the students' experience of Taiwan, the central districts of towns, the equivalent of Main Streets, are still quite populous, teeming with shops still catering to their residents' needs and open-air markets. Lowell's depiction of a desolate Main Street is indeed a contrast to students' expectations. As for exploring other emotional attachments to a small town like that described in the poem, hyperlinks could be made for Bob Seger, a popular singer of the 1970s and his hit song "Main Street." Likewise, the idea of the back porch brings connotations of an intimate vision of home. For "back porch," Internet searches yield a wide range from those belonging to upscale homes to those attached to smaller ones. The idea of the back porch where there is privacy for gatherings with family or friends in small towns may be a cultural concept needing a teacher's explanation.

The dwelling places, the architecture, of the place as alluded to in the poem are a special aspect of the journey through the Maine village setting. The dowager in a Victorian frame of mind, a term for the architecture and her exclusiveness, is buying up the "eyesores" that she has to look upon and allows them to run down. Real estate sites offer such older houses for sale in this area of New England, and students can examine the architecture of previous centuries in the form of advertisements of such houses for sale. As to be expected, the Internet searches are based on consumerism. These places on sale, unlike those in Lowell's example, have been decorated, and decorators' advertisements can also be easily accessed on hypertext. Church architecture brings the student into contact with a special aspect of the New England landscape. In the last two stanzas, like Bishop's "The Armadillo," which juxtaposes religion and the animals of nature, Lowell contrasts what he may see as the stark Puritanism of the New England Protestant religions with the fire in the sky lanterns at the Feast of St. John in Brazilian Catholicism in Bishop's poem. The students can view the Trinitarian Church and enter the church, as well as find out about its activities by accessing The Trinitarian Congregational Parish of Castine. It is a graceful white wooden building with a high steeple constructed in 1829 on Main Street. It is typical of the New England region.

The human characters in the poem are types. The "summer millionaire," who has sold his sailboat or yawl, is visible in the appearance he gets from the clothes he wears straight from a catalogue for sportswear from the prestigious L. L. Bean Company, still in existence and taking online orders. The sale of the nine-foot yawl associates to the figures of lobstermen, who bought the millionaire's boat. What was once for pleasure is now used for livelihood. Lobster fishing is a main occupation in this region. Links to this activity are quite informative whether they are professional, as the one titled "Maine Lobster Harvest" filmed by *America's Heartland*, or non-professional YouTube videos. Students interested in sailing can spend time comparing the boats pictured on Internet sites. The linkage to such sights brings images of consumerism in a capitalist country for the students in Taiwan to experience and evaluate.

Delving into the local folklore and customs of New England is not an easy task for the Taiwan student and may be quite shocking. The speaker drives his vintage Tudor Ford to a place overlooking the town graveyard. Internet searches for graveyards in Maine or New England in general bring students instantly face-to-face with the area's traditional "dark tourism."[3] There are YouTube cemetery tours in Maine and nearby states, including the place Elizabeth Bishop was buried, Hope Cemetery in Worcester, Massachusetts. Folk legends about ghost sightings, along with discussions of the paranormal, are recounted in videos evoking thoughts about mysterious occurrences. Students may be repelled by sites with invitations to visit particular graveyards and to watch performances of people dressing up as long-gone residents who are buried there. Even such virtual contact may be distasteful to them since visiting graveyards is a traditional taboo for any other purpose than ancestor worship on the Ching Ming Festival in spring when the family returns to pay respect to their ancestors at the burial site.

The subject of love is a focus of the poem through the images of "love cars" and a blues song. The "love car" could be described as a traditional cultural behavior of having a date with a lover in a car outside of town. In stanza 7 of the poem, there is an illusion to a traditional blues song titled "Careless Love." W. C. Handy, the father of American Blues, popularized the song, and there are a number of renditions on YouTube with varying lyrics that can be compared. A 1962 album by Ray Charles contains a particularly ironic version referring to the singer's blindness. Both here and in the poem, the song is a good example of Buell's "comic grotesque."

Finally, the focus turns to animals in this poem dedicated to Elizabeth Bishop, whose special focus on nature distinguishes her poems. To answer to the armadillo and the owls and the little rabbit who clenches his fist against the sky, Lowell brings the red fox, an object of hunters, and a brave family of skunks that ventures into town, even walking down Main Street and eating sour cream from his backyard. The fox is a more familiar animal to the students than the skunk, a likely hypertext target known for his black body with a white stripe and peculiar odor when frightened by human beings or other creatures.

In the virtual place created on the Internet through the introduction of the imagery, the teacher as hypertext designer or facilitator offers the possibility for at least some students to enter this language-created frame of mind. The navigation undertaken by the student through the limited choices of the Internet, however, can never replicate real life. It is merely a virtual experience of the place in one individual student's experience of the poems. Once the linguistic and cultural exploration is at an end, it is the task of the teacher to bring students back to the literary texts that led to the journey in the first place in order to obtain a holistic view. Navigating the poetic texts of Bishop and Lowell will have been a special experience the student will keep in her episodic memory.

In this model of teaching modernist poetry and acquainting students with other places and people's associations with place, we offer students poetry text enhanced with highlighted vocabulary and, with only some guidance, allow them do their own navigation through mostly self-selected links. As we ask them questions about their hypermedia journey, we encourage them to share their hypertext experience of the poems with each other.

For this model of interactive teaching of modern poetry we make the following observations with some certainty: First, the designer of hypertext for these poems is creating a space for navigation stimulated by two poems that exist in the environment of the other, both in literary history, as well as treatment of place and animal epiphany. Second, the from-top-to-bottom reading of the poems is changed into a complexly recursive one depending upon navigation strategies employed by different students. This is different from the recursive reading of poems presented only in textual form since the hypertext links lead the reader to other sites, which in turn can be followed to yet other sites, thus decentering the poem from the reader's mind. The interpretations expand and remain open. Third, even though the primary poetic texts open up to a large range of intertextual sites, the hypertext can counteract students' false assumptions, false cultural cognates, for images in the poems. Outstanding examples are of the deserted Main Street, USA, versus the main streets of Taiwan towns still teeming with people and of sky lanterns raised by fire at a religious festival in Brazil vs similar sky lanterns at festivals in Taiwan. Fourth, interacting with the objects in the space of the hypertext is a better heuristic experience of the poem for foreign language students than understanding the poem by means of the standard critical apparatus. But the understanding students obtain consists mainly of cultural background information that has to be evaluated as it is collected, and returning to reread the primary poetic texts is necessary for a holistic understanding. Fifth, the poems' negativity is threatened by the positive value of most Internet signs coming from the highly commercialized environment of the worldwide web. Examples were the L. L. Bean catalogue selling clothes for the outdoor sportsman, real estate advertisements for old houses in Maine, and both dark and conventional tourism of visits to graveyards, as well as itineraries for travel in Brazil and participation in its religious festivals. Sixth, the powerful juxtaposition of some images, like the graveyard and love cars, and the conflict between religion and nature in the closing epiphanies, may be able to recover the negative and critical tone, the resistance or defiance, in the poems. However, teachers may still need to bring the students' attention back to the primary text to bring their understanding closer to the poetic voices in the poems. And, finally, perceptive students will understand that sensitive poets are critically responding to their environments, the places where they dwell, and the places they love.

Locating images from poems in hypermedia is not only a journey to an "other" environment; it is a reflective experience, showing how we value the places in our lives. Through the language of poetry students learning foreign languages may access the denotations and begin to understand how words can have different values or connotations for different speakers. The use of hypertext further unearths a wealth of denotative and connotative meaning by allowing students to explore a wide range of language usage and make critical judgments about the hypertext links they find. The resulting montage of images draws students' attention to how we, either as distinct cultures or as individual human creatures, define our place in the world.

NOTES

1. The term "hypertext" refers to Internet navigation connections made to a primary text. Since not only reading, but listening and watching videos, can be accessed through links, the term has been replaced in most scholarship by "hypermedia." However, in our discussion with a literary focus on the images of poetry we have continued to use the former term.
2. The concept of "implied reader" derives from the reader's response theory developed by Wolfgang Iser and other critics. The theory is concerned about how the reader of books subjectively weaves together the text in response to the words of an author.
3. "Dark tourism" is a kind of tourism to places associated with death and disaster as defined by Lennon and Foley. Visits to graveyards have a special appeal for some tourists.

REFERENCES

Axelrod, Steven Gould, and Helen Deese, eds. *Robert Lowell: Essays on the Poetry*. Cambridge: Cambridge University Press, 1986. Print.

Balcytiene, A. "Exploring Individual Process of Knowledge Construction with Hypertext." *Instructional Science* 27(1999): 303–328. *Springer*. Web. 2 October 2015.

Berryman, John. "Despondency and Madness." Parkinson, *Robert Lowell: Essays on the Poetry*. 124–130.

Bishop, Elizabeth. *The Complete Poems.* New York: Farrar, Straus and Giroux, 1965. Print.

———, and Robert Lowell. *Words in Air: The Complete Correspondence of Elizabeth Bishop and Robert Lowell.* Ed. Thomas Travisano with Saskia Hamilton. Farrar, Straus & Giroux, 2008. Print.

Bolter, J. D. *Writing Space: the Computer, Hypertext, and the History of Writing.* Hillsdale, NJ: Lawrence Erlbaum, 1991. Print.

Bromme, Rainer, and Elmar Stahl. "Is a Hypertext a Book or a Space? The Impact of Different Introductory Metaphors on Hypertext Construction." *Computers and Education* 44 (2005): 115–133. *ResearchGate.* Web. 2. October 2015.

Buell, Lawrence. *The Future of Environmental Criticism: Environmental Crisis and Literary Imagination.* London: Blackwell, 2005. Print.

———. "*The Comic Grotesque.*" *New England Literary Culture: From Revolution through Renaissance.* Cambridge: Cambridge University Press, 1986. 335–350. Print.

Charles, Ray, perf. "Careless Love." *Modern Sounds in Country and Western Music.* Rec. 5–7 February 1962. Capital, 1962. LP. *YouTube.* 4 October 2015.

Crosby, Martha E., and Marie K. Iding. "The Influence of Cognitive Style on the Effectiveness of a Multi-Media Tutor." *Computer Assisted Language Learning* 10.4(1997):375–87. Print.

Dwight, J., and J. Garrison. "A Manifesto for Instructional Technology: Hyperpedagogy." *Teachers College Record* 105.5 (2003): 699–729. *EdITLib.* Web. 2 October 2015.

Eagleton, Maya B. "Making Text Come to Life on the Computer: Toward an Understanding of Hypermedia Literacy." *Reading Online* (July, 2002): 15–31. *EBSCOhost.* Web. 2 October 2015.

Gaggi, Silvio. *From Text to Hypertext: Decentering the Subject in Fiction, Film, the Visual Arts, and Electronic Media.* Philadelphia: University of Pennsylvania Press. 1997. Print.

Gardner, Howard. *Frames of Mind.* New York: Basic Books, 1993. Print.

———. *Intelligence Reframed: Multiple Intelligences for the 21st Century.* New York: Basic Books, 1999. Print.

Gilbert, Sandra M. "Mephistophilis in Maine: Rereading 'Skunk Hour.'" Axelrod and Deese, *Robert Lowell: Essays on the Poetry.* 70–79. Print.

Gioia, Dana. "Studying with Miss Bishop." Edited by George Monteiro. *Conversations with Elizabeth Bishop.* Jackson: University of Mississippi Press, 1996. 139–155. Print.

Goldensohn, Lorrie. *Elizabeth Bishop: The Biography of a Poet.* New York: Columbia University Press, 1992. Print.

Graff, Martin. "Individual Differences in Hypertext Browsing Strategies. *Behavior and Information Technology* 24.2 (2005): 93–99. *ResearchGate.* Web. 2 October 2015.

Haseltine, Patricia, and Tsui-lan Chen. "Reading the Images of Foreign Language Text: Energiea, Multiple Intelligence, and Hypertext." National Science Council, R.O.C. Unpublished Research Project, April, 2001. Report.

Iser, Wolfgang. *The Implied Reader.* Baltimore: The Johns Hopkins University Press, 1974. Print.

Kalstone, David. *Becoming a Poet: Elizabeth Bishop with Marianne Moore and Robert Lowell.* New York: Farrar, Straus, & Giroux, 1989. Print.

Kramer, Lawrence. "Freud and the Skunks." Axelrod and Deese, *Robert Lowell: Essays on the Poetry.* 80–98.

Laurens, Penelope, "'Old Correspondences': Prosodic Transformations in Elizabeth Bishop." *Elizabeth Bishop and Her Art.* Lloyd Schwartz and Sybil P. Estees. Ann Arbor: University of Michigan Press, 1983. Print.

Landow, George P. *Hypertext 3.0: Critical Theory and New Media in an Era of Globalization.* Baltimore: Johns Hopkins University Press, 2006. Print.

Lawless, Kimberly A., Robert Mills, and Scott W. Brown. "Children's HypertextNavigation Strategies." *Journal of Research on Technology in Education* 34.3 (Spring, 2003): 274–284. *ResearchGate*. Web. 2 October 2015.

Lennon, John, and Malcolm Foley, eds. *Dark Tourism: The Attraction of Death and Disaster*. London: Continuum, 2000. Print.

Lombardi, Marilyn May, ed. *Elizabeth Bishop: The Geography of Gender*. London: University of Virginia Press, 1993. Print.

Lowell, Robert. "On Skunk Hour." Parkinson, *Robert Lowell: A Collection of Critical Essays*. 131–134.

———. "Robert Lowell Reads 'Skunk Hour.'" *YouTube*. 17 August 2010. Web. 2 October 2015.

———. *Selected Poems*. New York: Farrar, Straus & Giroux, 1976. Print.

"Maine Lobster Harvest." John Lobertini. *America's Heartland*, Episode 517. Sacramento, CA: KVIE. Web. 4 October 2015.

Martz, William J. *The Achievement of Robert Lowell*. New York: Scott, Foresman, 1966. Print.

McGrath, Diane. "Hypertext, CAI, Paper, or Program Control: Do Learners Benefit from Choices?" *Journal of Research on Computing in Education* 24. 4 (1992): 513–532. *Tandfonline*. 2 October 2015.

Morgan, Wendy, and Richard Andrews. "City of Text? Metaphors for Hypertext in Literary Education." *Changing English* 6.1 (1999): 81–92. *Tandfonline*. 6 July 2006. Web. 2 October 2015.

Parkinson, Thomas, ed. *Robert Lowell: A Collection of Critical Essays*. Englewood Cliffs, NJ: Prentice-Hall, 1968. Print.

Raffel, Burton. *Robert Lowell*. New York: Frederick Ungar, 1981. Print.

Rudman, Mark. *Robert Lowell: An Introduction to the Poetry*. New York: Columbia University Press, 1983. Print.

Ryan, Marie-Laure. "Immersion vs. Interactivity: Virtual Reality and Literary Theory." *Postmodern Culture* 5.1 (1994): n.pag. *Project Muse*. Web. 20 February 2016.

Schreibman, Susan. "Computer-Mediated Texts and Textuality: Theory and Practice." *Computers and the Humanities* 36 (2002): 283–293. *JSTOR*. Web. 2 October 2015.

Seger, Bob, and The Silver Bullet Band, perf. "Mainstreet." *Night Moves*. Released 22 October 1976. Capitol, 1977. *YouTube*. 4 October 2015.

Spiro, R. J., P. J. Feltovitch, M. J. Jacobson, and R. L. Coulson. "Cognitive Flexibility, Constructivism and Hypertext: Random Access Instruction for Advanced Knowledge Acquisition in Ill-structured Domains." *Educational Technology* 31.5 (1991): 24–33. *ACADEMIA*. Web. 2 October 2015.

Thomas, Herbert. "The New Literacy?: The Challenges of Hypertextual Discourse." *Computer Assisted Language Learning* 10.5 (1997): 479–489. *Tandfonline*. 9 Jul 2006. Web. 2 Oct. 2015.

Travis, Molly. A. "Cybernetic Esthetics, Hypertext and the Future of Literature."*Mosaic* 29.4 (1996): par 3. *Questia*. Web. 2 October 2015.

Tripp, Steven, and Warren Roby. "The Effects of Various Information Resources on Learning from a Hypertext Bilingual Lexicon." *Journal of Research on Computing in Education*, 27.1 (1994):104. *Tandfonline*. 25 February 2014. Web. 2 October 2015.

Tuan, Yi-Fu. "Island Selves: Human Disconnectedness in a World of Interdependence." *The Geographical Review* 85.2 (1995): 229–230. *Questia*.Web. 2 October 2015.

———. *Space and Place: The Perspective of Experience*. Minneapolis: University of Minnesota Press, 1977. Print.

———. *Topophilia: A Study of Environmental Perception, Attitudes, and Values*. New York: Columbia University Press, 1990. Print.

Wen, Meichum Lydia Chin-Chung Tsai, Hung-Ming Lin, and Shih Chyeuh Chuang. "Cognitive-Metacognitive and Content Technical Aspects of Constructivist Internet-Based Learning Environments: A LISREL Analysis." *Computers and Education* 43 (2004): 237–248. Print.

III

DRAMA AND ETHOS BEYOND BORDERS

7

Shakespeare's *Cymbeline* in the College English Classroom during the Sunflower Student Movement

I-Chun Wang

There are various ways for college students to learn English in the college classrooms. In East Asian countries, students learn English through grammatical and listening drills, business or professional communication, and literature or liberal studies as well, depending on students' competence. In the highest level of College English classes at Kaohsiung Medical University, the core content includes lyrics, Shakespearean sonnets, poems by Robert Frost and Langston Hughes, short stories by William Faulkner, Earnest Hemingway, and Margaret Atwood, and selections from dramatic works such as *Julius Caesar* and *Death of a Salesman*. The reception of English and American literature has been excellent among students, but instructors find college English involved cultural studies challenging whenever voices for justice exist in the social atmosphere. During the period of the Sunflower Student Movement in 2014,[1] I unintentionally covered *Cymbeline*, a drama concerning shaping identity under the power of Rome, and this drama appealed to Taiwanese students who were eager to learn and express their interpretations of the drama by localizing *Cymbeline* with their experience and understanding, which were beyond the English language teaching especially. The appeal of Sunflower Movement was to monitor cross-strait agreements, and to prevent Taiwanese local businesses from falling into vulnerable positions because of the unbalanced trade pact with China. During the three-week study on *Cymbeline*, this Shakespearean play was localized and the classroom became a site to discuss Cymbeline's positioning and Taiwan's identity politics.

When teaching Roman plays, I oftentimes start with the definition of masculinity and military valor promoted by Roman culture, the relationship between the Britons and Roman imperial power, or the spatial concerns that stem from the expanding territory of Rome. In the introductory section, I covered *Antony and Cleopatra* and Fletcher's *Bonduca* as examples, both of which demonstrate ethnic groups on the borders of their empire. Although Shakespeare's *Cymbeline* and Fletcher's *Bonduca* are both topos representing how ethnic groups in Britain mediated or reacted against the Roman invaders, compared to Fletcher's *Bonduca*, a warrior queen of the Icenic tribe in defending her land, Cymbeline seems to be an ambiguous figure who finds it difficult to face aggressive Roman conquerors. However, the warrior queen witnesses her people defeated by Rome. *Cymbeline* is a play written during the period when King James tried to unite England and Scotland; the characters' shifting of identity may suggest anxieties and political tensions in early modern Britain. However, stressing geographical and spatial concerns in teaching, one cannot easily help Taiwanese students comprehend Shakespeare's *Cymbeline* without touching upon the formation of British identity. In this sense, the long British *history of* political compromise under the Roman rule, the phenomenon of lacking a shared identity among the Britons, as well as the limits and obstacles for identity formation are all necessary information for students. A few students who were aware of Caesar's Gallic War across the strait and the impending danger of Britons brought up issues as related to military tensions between China and Taiwan in their response notes and group presentations. These students were more interested in relating *Cymbeline* to Taiwan-China relations, basing on their own experience as citizens with Taiwan consciousness and the power struggle across the Taiwan Strait. The College English class therefore turned out to be a class for localizing Shakespeare's *Cymbeline*.

Localizing Shakespeare has been a common practice on the modern stage; each new version of Shakespearean plays with localized implications justifies reader's response. Situating geopolitics in this Shakespearean play with Taiwan experience was not unexpected especially during the time period of the Sunflower Movement. Most students understood *Cymbeline* from the perspective of British literature, but quite a few students wrote in their response notes or their oral presentations, relating that *Cymbeline* mirrors up Taiwan's political situation. In the class of 2014, quite a few students not only analogized captured students in the Sunflower Movement to literary characters such as Prometheus and Caliban, and examined the value of freedom as implied in *Cymbeline*, but also questioned the use of violence by the government against democratic voices during the protest.

The Sunflower Movement occurred during the spring semester of 2014, when 400 students occupied the Legislative Yuan (the Legislative Ministry) to protest the Service Trade Agreement with China, while almost half a million people joined the protests in Taipei city.[2] During these grand-scaled protesting periods, thousands of college and high school students joined, protesting at various sites, and the priority for most instructors was to keep students in classrooms by responding to students' questions and listening to their voices as related to their civil protests. Some students were in tears and some agitated when the Taiwan issue was brought in by student presenters. Taiwan has tangled historical, cultural, and economic ties with China; whenever China claims reunification, resistance power appears in TV news and programs in Taiwan. During the Sun Flower Movement in 2014, College English classrooms became sites not merely for instructors to perform instructional tasks, but also for analysis, evaluation, and interpretation of *Cymbeline* via reasoning Taiwan-China issue.

I.
BORDER POLITICS IN COLLEGE ENGLISH CLASSROOM

College English Level 3 in Kaohsiung Medical University is a course conducted in English. Teaching *Cymbeline* in the three two-hour sessions, I oftentimes use border politics to explicate the imperialist expansion of Rome, after we go through selected episodes and commentaries from critics. Students are required to respond to the themes, raise questions, or provide interpretations, and some students choose *Cymbeline* as a focus to do presentations or perform mini-plays. Five episodes required to read include Act I, Scene iv (Rome, Philario's house), Act III, Scene i (Britain, A hall in Cymbeline's palace), Act III, Scene iii (Wales, a mountainous country with a cave), Act V, Scene ii (Field of Battle between the British and Roman camps) as well as Act V, Scene v (Cymbeline's tent). Most students were impressed by Shakespeare's sophisticated use of historical material in representing the border issue, and appreciated the emergence of autonomy of identity in early modern Europe. The framework and romantic plotline of *Cymbeline*, such as Cymbeline's beautiful daughter Imogen and her exiled husband Posthumus, Posthumus's relation with Rome, and the political encounter between the Roman and Briton camps are all appealing to Taiwanese students. The former plotline develops from Posthumus's love test via his agreement with Iachimo's attempt to try Imogen's innocence, and the family reunion of Cymbeline and his two abducted sons. The second involves the impact of Roman culture upon Britain in the case of Cymbeline and Posthumus's father, and the last one is interlaced with Roman imperialism, resistance in Britain as well as the possibility of political autonomy.

The questions focusing on *Cymbeline* includes Cymbeline's role as king, which is complicated, first by his rejection to pay tribute to Rome, his attempting to defend his border, and later by his choosing to be a client king. Cymbeline's decision to be a client king may signify a possibly harmonious relationship between Rome and Britain, but it may also suggest his necessary political concern in submission to a power much greater than his own. Some students suggested Cymbeline himself, being Romanized, has already lost part of the local identity, while some prefer to suggest Cymbeline makes his final solution purely for peace. During the three sessions or six hours' class, we covered recent scholarships in cultural studies that tend to take *Cymbeline* as an example of self-justification as a nation (Lockey 115; Redmond 170). Scholars' comments for discussion, for example, Jodi Mikalachi's theory of historical continuity as one way to understand Britain's past (302), and Mikalachki's and Alison V. Scott's views that this play is primarily an example of early modern nationalism, were listed for discussion. Although to historians such as R. G. Collingwood and J. N. L. Myres, King Cymbeline (also known as Cunobelinus or Kymbelinus) might have never sought the friendship of Rome (74–84), yet in Raphael Holinshed (1529–1580), during the reign of Emperor Claudius (10 BC–54 AD), Roman armies arrived (Book III, ch. xviii), Britain was disabled by the Roman military conquest and Plautius, a Roman, became the first governor of Britannia (Collingwood

74–84), and Cunobelinus's descendents became subordinated to Rome. In AD 122, to fend off the barbarians from the north, the governor ordered the construction of Hadrian's Wall with an aim to guard his realm, and the rule of Rome over Britain lasted until the fifth century.

Students are interested in border politics. Geopolitical functions of Hadrian's Wall (also called Pick's Wall) can be easily understood via the Great Wall of China. The main function of both the Great Wall of China and Hadrian Wall in Britain was to fend off "the uncivilized" nomadic tribes from the north. The differentiation of self and the other was emphasized in students' presentations, but quite of few acknowledged the multiethnic culture phenomenon foreshadows the assimilation of Cymbeline to Roman culture. Some found the trait of multiethnic culture in Ancient Britain echoes the environment when some indigenous tribes in Taiwan were trying to maintain their tribal safety by withdrawing themselves to mountainous areas, while the other tribes were assimilated and their lands stripped of. Indeed, as Hadrian Wall, northwest frontier of the Roman Empire, served the borderline between the Romans and the barbarians. It also implied the loss of autonomy and assimilation policy that had brought all British political leaders under the gaze of Rome.

The issue as related to democracy and autonomy is a fervent topic in the colleges in Hong Kong and Taiwan. The Umbrella Revolution (2014 Protest) in Hong Kong voiced political autonomy, while the Sunflower Protest in Taiwan that occurred half a year earlier was against "Cross-Strait Agreement on Trade in Services". We were supposed to discuss the topics such as how Cymbeline keeps his sovereignty within his own borders, the crisis Cymbeline encounters, and the possibility of formation of British identity through creolization. Intermittently, students tried to explore identity politics in *Cymbeline*, by using Taiwan issue as a paratext. Indeed, after KMT (Kuomingtang or Nationalists) retreated to Taiwan in 1949, Taiwan Strait has become the border between China and Taiwan. The United States officially recognized China in 1979, which overwhelmed Taiwanese people for its diplomatic decision. Although in the past years the United States still recognizes a policy of maintaining essential parity across Taiwan Strait region through arms sales to Taiwan, unofficial trade and investment relations, China has never promised to defer the use of force to settle the Taiwan problem. Over 60 percent of Taiwanese take Taiwan as a sovereign state, but China claims that the key *issue* regarding *the use of force* is whether it is an independent nation-state (Kang 93). In students' explication, geography in *Cymbeline* involves cultural discrepancy and confluence between Britain and Rome; however, teaching *Cymbeline* with concerns of Taiwan issue has unavoidably become an on-going phenomenon when trading issue was in turmoil.

The cultural relationship between Rome and Britain is eminent in terms of characterization. Shakespeare's *Cymbeline* derives from the legendary history of a British king Cunobelinus (also spelled as Cymbeline, Kymbeline). Cymbeline was brought up in a Roman palace as a hostage (Bulfinch 26) when his father was a client king of Rome. It was said that when Cymbeline was in Rome, he was converted to Christianity by the apostle Paul (King 65–71). The historic Cymbeline had a son, Adminius, who was banished, but the prince fell under the protection of Roman Emperor Caligula (reigned 37–41 AD). The identity issue in *Cymbeline* is highlighted through the contrasts between Cymbeline and his queen; she not only insists on political autonomy but also recognizes valor in Britons' national character. Shakespeare, however, represents the queen as a wicked wife who plots to get Cloten, her son by her former husband, to succeed to the throne. Her scheme includes trying to get his son married to Princess Imogen, without knowing Imogen has secretly married Posthumus Leonatus. Because of this, Posthumus escaped to Rome where his father's old friend Philario acommodates him, while Imogen, disguising herself as an English emissary to Wales, self-consciously seeks to anglicize the native culture. In the story, Cymbeline's two sons, Guiderius and Arviragus, were long ago kidnapped by a banished lord, Belarius, who, out of a sense of guilt, brought up, in the simplest life style, the two princes, residing on the border of England and Wales. The play ends with a family reunion, the death of the wicked queen and Cymbeline's acknowledgement of himself as a client king of the Roman Empire in front of the Roman ambassador. As the queen and her son insist that the Britons cannot be subjugated, they represent voices of political autonomy and antagonism toward Rome. It is significant to stress the background of *Cymbeline* is the constant tension of war. Julius Caesar's conquest of Gaul was completed by 51 BC, and the territory of Rome was consequently extended to the English Channel and the Rhine (Frere 17); in the next few years, Caesar turned his attention to Britain and a number of British tribes were glad to receive Roman support. In Shakespeare, voices of autonomy comes from the Queen and her son Cloten who stresses the self-sufficiency of the Britons: "Britain is a world by itself" (III. i. 18), while groaning, "Why should we pay tribute? If Caesar can hide the sun from us with a blanket or put the moon in the pocket, we will pay him tribute for light" (III.i. 40–3). In a similar tone, the queen confirms that Britain, being geographically separated from other lands, has developed itself into an independent self-sufficient kingdom. In this way, she daringly challenges the power of Rome by sneering at Caesar as a defeated conqueror in his two attempts at the conquest of Britain. Shakespeare's Cymbeline, empowered by his wife's insinuating voice, agrees that the Britons were free before Caesar stretched his power and the later oppressors imposed the yoke upon his people. Manifestly, Cymbeline has stopped paying tribute for three years. Warning Cymbeline about his attitude, the Roman ambassador conveys that "war and confusion" will soon take place. Since he has chosen to remain independent, confirming with equally threatening words that his men have become "warlike people," ready to fight (III. i. 57–82), with a network of resistance. As the ambassador realizes that his trip to Britain to exact tribute has been futile, he decides to leave, asking for a retinue to accompany him to Milford Haven, which is the border town close to Wales. Right after he provides the hostile ambassador with a military escort to the border, Cymbeline commands his troops to prepare chariots and horsemen for war. Although Cymbeline eventfully decided to be Rome's puppet king, not till the final scene of the play does Cymbeline seem to have realized the senselessness of the political crisis and reveal his intention to be allied with Rome by negating the queen's previous acclamation of Britain's autonomy.

The border politics in *Cymbeline* is reinforced by geographical concerns. Geographical borders spanning mountains, rivers, lakes tend to carry political implications. In this drama, rivers are potent political boundaries. While the River Trent marks a border between the north and south of England, the River Severn, the longest river in Britain, which flows from the Cambrian Mountains and discharges into the Bristol Channel, forms the border between England and Wales for much of its course, but it also serves as a natural boundary between the Roman settlement ruled by Aulus Plautius and the subordinated Wales (Sanders 45). The Severn River helps define Cymbeline's political borders and at the same time implies the subtle Anglo-Wales relationship, because during the early modern period, Wales was destined to be ruled by the king of England (Schwyzer 1). Shakespeare's Milford Haven, situated in the south of Pembrokeshire, on the border of England and Wales, signifies a place without a distinct cultural identity. As Ronald Boling notes that Milford Haven, commonly called "Little England beyond Wales," is like an English colony (34). Willy Maley also contends that Milford Haven represents "a little England" (127–28), a weak spot for defense (Dudly 143). The queen envisions her ideal kingdom as free; her national pride rejects the new identity imposed by the invading Romans, but Britain's worthiness as a self-contained island is not available during the reign of Cymbeline because skirmishes and war have started. Even though Cymbeline captured Lucius, returning legions led by Lucius may signify further serious danger and it becomes a pivotal point for Cymbeline's final resolution. Although David Hughes notes the historical Cymbeline in Holinshed was one of the most valiant kings in Britain (82), Shakespeare's Cymbeline looks more like a plotter who chooses to reconcile with Rome because he foresees the conflicts would last indefinitely. Eventually the epic battle ends with Posthumus's commentary that a seeming peace is found, and there follows the Britons' negotiations with Lucius, who reappears at Cymbeline's palace, witnessing Cymbeline's vouching of further commitment to a harmonious relationship with Rome, while Cymbeline blames his "wicked queen" whose rhetoric had maneuvered him into war with the Romans.

II.
LOCALIZING SHAKESPEARE'S *CYMBELINE* IN ENGLISH CLASSROOM

Cymbeline is one of the most ambitious and complicated Shakespearean plays, and a long-neglected one. With the trends of localization of Shakespeare in various countries, the reception of *Cymbeline* in Taiwan's classroom manifests the concerns toward China/Taiwan relations. In response notes, group presentations and mini play performance, several questions were raised:

1. Can Cymbeline foresee the colonization by Rome from AD 43 to 409?
2. Is Cymbeline's story considered as an inglorious past of Britain?
3. Does the flowing of culture demise Cymbeline's national identity?
4. For what purposes did Cymbeline choose to be a client king?
5. Is there no other way to curb an imperialist like Rome or China?

The final act of *Cymbeline* seems to be the part for most discussions: General Lucius was defeated and captured, but Cymbeline is reconciled to Rome. To most students, Cymbeline's capturing of the Roman Lucius, which signifies victory, should be a separate matter from reconciliation. Cymbeline's making peace with the Romans led to debates in class. Some students tend to suggest Cymbeline should have taken hostages of the leaders of Roman legions and some suggest peace negotiations with Roman soldiers as hostages since the presence of imperialist Rome could not last long. Taiwanese students are more sensitive toward the border issue and the interpretation of political autonomy. To some students, indeed, analogies between Taiwan and Ancient Britain involve not only the issues of political autonomy, imperialist encounter, and border politics, but also the geographical locations of Britain and Taiwan. In the mini-play invented by a group of students, Cymbeline's queen escapes, but comes back with the other tribal leaders in Britain to defend the border. Students' masculine autonomy with unification of tribal groups seems to envision a powerful leader to guarantee the safety of the realm of Britons. The mini-play was considered as an allegory, a revised romance or a fantasy that invaders will find their doom. It is understandable that yearning for autonomy and democracy among college students implies resistance against possible war across the Taiwan Strait. Students even raised the question why Shakespeare fictionalized Cymbeline's queen as wicked. The other questions from students include cultural discrepancy among England, Wales, and Rome, and stresses the cruelty and unpolished ways of the Britons and the civility of Rome, especially how they treat their captives as well as how the Britons fight for their identity. Indeed, after KMT retreated to Taiwan in 1949, the Taiwan Strait has become the border between China and Taiwan. Within the island, political autonomy has been practiced but the threat across the strait has been an on-going problem. Within the class, while sensing the crises across the Taiwan Strait and across the border of Cymbeline, students generally note nation building in different periods of British and Taiwanese histories, as well as geographical politics as related to collective identity. As a student from Singapore mentioned, without collective identity, any single state would have its problems and identity crisis; his examples included Southeast Asian countries under the power of imperialist Japan. A student from Hong Kong compared border politics and power struggle by stating that identity formation happens when people experience in real life, and he added, domineering culture and authorities cause resistance.

For the topics like cultural assimilation, and cultural confluence, an analogy between Britain and Taiwan was raised. It is important to know that in *Cymbeline*,

Britons have already been Romanized, with Cymbeline knighted by Augustus and the Britons worshipping Jupiter and the other Roman gods. Shakespeare's ancient Britain is a realm constantly threatened by the Romans, and the fact that Cymbeline had once been a client king complicates the issue of cultural identity. Students found that when the queen suggests the strait separates the Romans off the shore of Britain, she might have neglected the fact that the border on Milford Haven would become a rift to crash Cymbeline's newly established sense of autonomy. Furthermore, one similarity between ancient Britain and Taiwan lies in cultural confluence, the former with Rome and the latter, China.

In *Cymbeline*, cultural discrepancy involves cultural geography. Referring to the Britons as people of the wilderness, and Cymbeline's willingness to accept cultural assimilation by Rome, Eric Heinze expounds the cultural standards in this play that in the eyes of Rome, "Britain stands as does Wales in the eyes of Shakespeare's Britain" (373). In a similar way, Shakespeare's contemporary historian William Camden suggests "the yoke of the Romanes . . . chased away all savage barbarism from the Britons minds" (63). Now that the Romans in a way "tamed" the ancient British landscape to suit their military and other purposes (Sullivan 21), it is understandable that Cymbeline could only hold a vision for his kingdom based on the memories that he retained of Roman culture and a kingdom without the repeated threat of impending war. Ironically, colonization by Rome eventually prevailed and bloodshed was seen in numerous tribal groups represented in plays such as *Bonduca*, *Zenobia* as well as *Caratacus*. Students took this part of *Cymbeline* as an allegory, since emergence of any super power, according to them, may develop itself as an impending threat. Students were all aware of the impact of colonial powers on Asian countries and the catastrophes created by the Imperial Japan. As students suggest, to balance power and *prevent* the outbreak of another war in Asia is a diplomatically sensitive but significant subject.

III.
READING CULTURAL AND POLITICAL CRISIS IN *CYMBELINE*

Shakespeare's *Cymbeline* is a play about the crisis encountered by ancient Britain and the final resolution made by her king. The crisis includes the slip of identity formation, military invasion of the Romans, as well as the modification of Britain's political identity. At the end of the play, Cymbeline's troops fight and win the battle but Shakespeare lays bare that identity can be multilayered, both across the border and on the battlefield. Imogen's exiled husband Posthumus, for example, finds refuge in Rome; eventually he is seen fighting with the Roman army against Britain. His shift back to his British identity is represented in the last part of the play, when he dresses himself as a peasant, fighting for England: "I'll disrobe me / Of these Italian weeds, and suit myself / As does a Britain peasant; so I'll fight / Against the part I come with" (V. i.). This shifting of identity not only signifies Posthumus's blurred cultural identity but also parallels Cymbeline's alliance with Rome at the end of the play for tentative peace. Students in the class were curious to find that even though Roman legions lost the battle, the rhetoric of empire represented by Lucius still persists, and the reign of the Romans over Britain lasted for more than three centuries. The students even questioned if a person with blurred identity, like Cymbeline, tends to be more easily tamed.

The issue of multilayered cultural identity is a significant question raised in class. Since nationalism involves boundaries and space, does the blurred boundary devalue the sense of solidarity? *Cymbeline* is one of the Shakespearean plays that show multifarious aspects of imperialism and voices of resistance. Rome is Cymbeline's second home, where he was brought up, educated and knighted. Cymbeline stresses that Rome was the place where he received honor. For the injustice and tribute imposed by the Romans, he tried to stage rebellion but alliance with Rome became his resolution. Alexander Leggate finds that Cymbeline and Caius Lucius share mutual esteem (278–82). By positioning Cymbeline in a shifting position in terms of identity, this Shakespearean drama represents a complicated situation when early modern European nations were competing with each other for territories and for positioning of people, this play reflects not only early modern unease in defining pre-Roman British identity but also the concern over cultural diversity and mobility on the edges of the Roman empire (Mikalachi 303).

Colonization involves a process of cultural assimilation. Francis Haverfield notes the process of Romanization in its provinces lies in the fact that "they were given a civilization" (quoted in Webster 209). Through accuralization, and cultural interchange, the invaded somehow lose their sense of vigilance. To some students in the class of College English, blurred identities may suggest tentative peace and mutual recognition, but once antagonism on one side occurs, there would be no sufficient time for the other side to react. From the perspective of sociology, anxieties over any oppositional groups are due to non-trust and political unease in the long-term cultural memories. When the Romans practice stable regimes to insure peace and security along the borders, Lucius, with persuasive tone, exemplifies amicability, and Cymbeline the winner of the battle affirms his friendship. Students found that Lucius never shows any further attempt to help consolidate Roman conquest and alliance, but Cymbeline seems to understand alliance with Rome is almost inescapable.

To interpret Shakespeare from the perspective of post-colonialism or localization has been a trend over the past ten years, and it is important to explore political subtlety in the relationship between the Roman legions and the client kings of Britain. Shakespeare's *Cymbeline* is set in a Britain that was not yet united and was under the threat of a powerful empire and native tribalism. Similar to Prasatugas, King of the Iceni, Cymbeline (Kymbeline) in history also signed a peace agreement in accord with Rome (Milton 367). Though Cymbeline has tentatively defeated Lucius' troops, he knows the Britons will still need to confront overpowering Roman forces beyond his borders. Diplomacy contact and trade relations were common practices for Rome before attempting conquest but chances were that deprivation of autonomy was an inescapable result. Paralleling Prasatugas's invincible queen Boudica, Caractacus was a leader who resisted Roman for some years but was eventually subdued and brought to Rome in chains. His stunning speech once moved Claudius to take off his chains: "If you Romans choose to lord it over the world, does it follow that the world is to accept slavery?" (Tacitus 152). The motif of resistance and nativist counterattack against colonizers were stressed by another group of students; their study on the films, *Warriors of the Rainbow: Seediq Bale* and *Boudicca: Warrior Queen of Ancient Britain* (2003) not only represented colonial rule in Britain and Taiwan, but also contrasted the warrior queen's soldiers and Seediq warriors who fought against the Romans and Japanese. In historical records, Seediq women ended their lives before warriors went to the battlefield, while Boudicca's warriors who valiantly defended their land were massacred.

IV. CONCLUSION

Cymbeline is a play representing a history of violence and cultural memory of resistance and reconciliation. In the class of College English, the approach to *Cymbeline* based on Taiwan-China relationship was demonstrated. Students tend to see Rome in *Cymbeline* as a prototype of imperialism that uses military force and colonization to increase territories and benefits. Most students could parallel the alliance between Rome and Britain to multiple relations among England, Scotland, and Ireland, while several students perceived that, by stretching across the Severn, and anglicizing Milford Haven as Little England, Cymbeline's power and his small-scaled empire are in formation. It was agreed in class that when Cymbeline's realm was threatened by the in-coming Roman legions, Cymbeline chooses glorification of Roman heritage, creolizing Britain with Roman culture not merely to find peace with the Roman Empire but to empower himself among the indigenous tribes. In this sense, the British were not all innocent victims of their own subjugation (Creighton 2). To the aging king, the alliance with Rome signifies not only peace but also the rejuvenation of Britain (Moffet 217).

The Sunflower Movement is a movement to protect Taiwan's trade benefit and democratic practice. In College English classes, we professors rarely touched upon Taiwan consciousness or Taiwan issue; during the period of the Sunflower Student Movement, each professor needed to accommodate various questions. Compared to the 2012 classes of the graduate program at National Sun Yat-sen University where a comparative study of Milton, Spenser, and Shakespeare in terms of the issue of cultural identity helps students understand cultural-political significance of the play, students at College English class chose to question identity issues with their own life experience. In the atmosphere affected by the Sunflower Movement, localization of Shakespeare's *Cymbeline* in College English class demonstrates border and identity, justice and benefit, as well as political benefits for ancient Britain and Taiwan. At the conclusive moment for *Cymbeline*, a student affirmed that antagonism is not a good solution but Taiwan's benefit has to be maintained by the government when the delegates sign agreement.

Shakespeare's *Cymbeline* represents a kind of historiography of British identity, by depicting Cymbeline's era as one at a crossroads of British history. For college learners in the classroom of English, students discussed Taiwan's position at crossroads. As students found Shakespeare's aging Cymbeline looks into the bright side of cultural integration and assimilation, they would rather envision priority on Taiwan's benefit. Not surprisingly, *Cymbeline* as a masterpiece that embodies wisdom and depths indirectly induced those voices that had never been heard in my English classes—the care for Taiwan's future. Affected by Sunflower Student Movement, the College English classroom unprecedentedly witnessed localizing Shakespeare's *Cymbeline* from the perspective of Taiwan experience and learning language through the voices for peace and democracy.

NOTES

1. The Sunflower Student Movement refers to a protest movement in Taiwan driven by a coalition of students and civic groups against the Cross-Strait Service Trade Agreement (CSSTA). From March 18 to April 10, 2014, hundreds of students occupied the "Legislative Yuan," and, later, also the Executive Yuan of the Republic of China (Taiwan): "The activists protested the passing of the Cross-Strait Service Trade Agreement (CSSTA) by the ruling party Kuomintang (KMT) at the legislature without clause-by-clause review. However related movements are ongoing, in particular the Ministry of Education (Taiwan) was stormed by Anti Black Box Movement protesters."

2. The number of participants reached half a million on March 30, 2014. See "The Black Tide of Sun Flower Movement Writes History." See Apple Daily, March 31, 2014. http://www.appledaily.com.tw/appledaily/article/headline/20140331/35736222/. Data retrieved on September 1, 2015.

REFERENCES

"Black Tide of Sun Flower Movement Writes History." *Apple Daily*, March 31, 2014. http://www.appledaily.com.tw/appledaily/article/headline/20140331/35736222/. September 1, 2015.

Boling, Ronald. "Anglo-Welsh Relations in *Cymbeline*." *Shakespeare Quarterly* 51.1 (2000): 33–66. Print.

Bulfinch, Thomas. *Bulfinch's Mythology: The Age of Chivalry*. Middlesex: The Echo Library, 2006. Print.

Camden, William. *Britain, or A Chorographicall Description of the Most Flourishing Kingdomes, England, Scotland, and Ireland*. Trans. Philemon Holland. London: George Bishop & John Norton, 1610. Print.

Collingwood, Robin George and John Nowell Linton Myres. *Roman Britain and the English Settlements: A History of English History from the Roman to Anglo Saxon Period*. New York: Biblio and Tennen, 1998. Print.

Creighton, John. *Britannia: The Creation of a Roman Province*. Oxford: Routledge, 2006. Print.

Dudley, G. "The Development of Milford Haven as a Major Port." *Journal of Navigation* 29 (1976): 141–69. Print.

Frere, Sheppard Sunderland. *Britannia: A History of Roman Britain*. London: The Folio Society, 1987. Print.

Heinze, Eric. "Imperialism and Nationalism in Early Modernity: The 'Cosmopolitan' and the 'Provincial' in Shakespeare's *Cymbeline*." *Social & Legal Studies* September 18 (2009): 373–396. Print.

Holinshed, Raphael. *The Historie of England, From The Time That It Was First Inhabited, Vntill The Time That It Was Last Conquered,* (An Electronic Edition) Gutenberg Project: http://www.gutenberg.org/files/16511/16511-h/16511-h.htm. April 5, 2014.

Hughes, David. *The British Chronicles, Book One*. Westminster, Maryland: Heritage Books, 2007. Print.

Kang, David C. *China Rising: Peace, Power, and Order in East Asia*. New York: Columbia University Press, 2007.

King, Ros. *Cymbeline: Constructin of Britain*. Aldershot: Ashgate, 2005. Print.

Knight, Charles. "Introductory Notice." *The Comedies, Histories, Tragedies and Poems of William Shakespeare*. London: Charles Knight. 1842. 189–95. Print.

Leggate, Alexander. "Canada, Negative Capability and *Cymbeline*." *Shakespeare in Canada: A World Elsewhere*. Toronto: University of Toronto Press, 2002. 274–91. Print.

Lockey, Brian. "Roman Conquest and English Legal Identity in *Cymbeline*." *Journal for Early Modern Cultural Studies* 3.1 (2003): 113–46. Print.

Maley, Willy. "*Cymbeline*, the Font of History, and the Matter of Britain." *Alternative Shakespeares 3*. Edited by Diana Henderson. London: Routledge, 2007. 119–37. Print.

Mikalachi, Jodi. "*Cymbeline* and the Masculine Romance of Roman Britain." *The Shakespeare Quarterly* 46. 3 (Autumn, 1995): 301–22. Print.

Milton, John. *The History of Britain: That Port Especially Now Called England*. Book One. London: Printed for R. Wilks, 1818.

Moffet, Robin. "*Cymbeline* and the Nativity." *Shakespeare Quarterly* 13.2 (1962): 207–18. Print.

Redmond, Michael J. "'No More a Britain': James I, Jachimo, and the Politics of Xenophobia in Cymbeline." *Shakespeare, Politics, and Italy: Intertextuality on the Jacobean Stage*. Surrey: Ashgate, 2009. 169–204. Print.

Sanders, Julie. *The Cultural Geography of Early Modern Drama, 1620–50*. Cambridge: Cambridge University Press, 2011. Print.

Scott, Alison V. "Imogen's fides: Secrets and "Good Faith" in *Cymbeline*." *"Rapt in Secret Studies": Emerging Shakespeares*. Edited by Danyl Chalk and Laurie Johnson. Cambridge: Cambridge Scholars, 2010. 133–53. Print.

Schwyzer, Philip. "A Map of Greater Cambria." *Early Modern Literary Studies* 4.2. Special Issue 3 (September, 1998): 4.1–13. http://purl.oclc.org/emls/04-2/schwamap.htm.

Shakespeare, William. *Cymbeline*. Edited by Martin Butler. Cambridge: Cambridge University Press, 2005. Print.

Sullivan, Garret. "Civilising Wales: *Cymbeline*, Roads and the Landscapes of Early Modern Britain." *Early Modern Literary Studies* 4.2. Special Issue 3 (1998): 1–34. Print.

"Sunflower Student Movement." https://en.wikipedia.org/wiki/Sunflower_Student_Movement, September 1, 2015.

Tacitus, Cornelius. *The Annals of Imperial Rome*. New York: Digireads.com, 2014.

Webster, Jane. "Creolizing the Roman Provinces." *American Journal of Archaeology* 105. 2 (April 2001): 209–25. Print.

8

MOOC "Global/Local Shakespeare"

New Approaches to Teaching Shakespeare in Taiwan and Beyond

Yilin Chen

Shakespeare becomes transnational and circulates rapidly through social media and the Internet in this digital age. These technologies facilitate a new form of online learning, MOOCs (Massive Online Open Courses), by which learners can have access to digital learning contents for free from anywhere. Following this idea, the Taiwan Ministry of Education (MOE) has funded MOOC "Global/Local Shakespeare," a course I designed to teach Shakespeare in a global context with an awareness of local Taiwanese culture. Taught by interactive online videos, this MOOC course provided a platform to study Shakespeare beyond English language and culture.

In the history of English teaching in Taiwan, the significance of "Global/Local Shakespeare" was its introduction to the ongoing globalization and localization of Shakespeare in Taiwan and other nations. Launched in September 2014, the course introduced performances and adaptations in different forms of popular culture from all over the world to examine and analyze Shakespeare's texts. Shakespeare was first adapted into *manga*, the Japanese comics, in the 1970s. By the twenty-first century, Shakespeare was published in numerous graphic novels and *manga* adaptations on the market. This course used diverse Shakespearean interpretations in *manga* adaptations and performances across the world to subvert the traditional impression of Shakespeare as a British canonical writer.

The course incorporated the transnational dissemination of *manga* Shakespeare into teaching, and examples were given to explain the glocalization of Shakespeare. *Manga* became a new medium to interpret Shakespeare's text, and it represented not only the consequence of cultural exchanges between the East and the West, but was also a testimony to the fact that the globalization of Shakespeare was not necessarily Anglophone Westernization. In the early twenty-first century, Japanese *manga* won popularity in the United Kingdom and inspired the British publisher SelfMadeHero to conceive the *Manga Shakespeare* series (Hayley 267). The British *Manga Shakespeare: Hamlet* (2007) was published around the same time when a Taiwanese *manga* adaptation of *Hamlet* (2006) came out. Both adaptations were categorized as *shôjo manga*, which literally means "girls' comics," to aim at a teenage female readership.

However, the globalization of Shakespeare began in the Age of Discovery, the impetus to the European colonial exploration and exploitation. With this in mind, the course I designed introduced localization of Shakespeare in Taiwan after Japanese rule (1895–1945). It emphasized the ways in which Shakespeare was negotiated in particular local cultures, especially the wave of popularizing and localizing Shakespeare in Taiwan from the 1990s onward. Shakespeare became the means to celebrate the newly claimed freedom and artistic expression of cultural diversity after democratization.

As the English language becomes the *lingua franca,* and the world is globally connected via Internet these days, this online course helped learners gain an international perspective on how Shakespeare has been related to each culture. By studying different Shakespearean adaptations in *manga* and performances, it drew attention to the interaction of Shakespeare and local cultures rather than treating Shakespeare as a foreign canonical object of learning. These new approaches also helped Shakespeare to become relevant on a personal level to people from different cultural backgrounds.

TRANSNATIONAL *MANGA* SHAKESPEARE

The MOOC Global/Local Shakespeare course used Japanese comics, *manga*, to teach Shakespeare. In Japan, Shakespeare has been incorporated into *manga*, *anime*, and computer games over decades. The rise of Japanese popular culture challenged the idea that globalization was primarily the spread of westernization. With the worldwide prevalence of Japanese popular culture, a new transnational trend of drawing Shakespeare in *manga* style appeared. Artists from different cultures imitated the *manga* techniques and assimilated them into their drawings. Consequently, the emergence of *manga* Shakespeare versions in different regions was a proof of the Japanese term, glocalization. These regional artists adapted Shakespeare in *manga* to fit their own localities or cultures in which their drawings were sold. This course compared the Taiwanese *shôjo manga* series of *Hamlet* (2006) by Meng Chen[1] and the British *Manga Shakespeare: Hamlet* (2007) by Emma Vieceli to analyze representations of Gertrude's sexuality in different cultures.

Given that *manga* became a new way of adapting Shakespeare's text, this course briefly introduced Japanese *manga* style and *shôjo manga*. In *shôjo manga* conventions, stories were primarily associated with spiritual love, Prince Charming, and boy love[2] because most of its readers were adolescent girls (Shamoon 104–105). As long as the romance was not concerned with heterosexual sex, the undefined sexually oriented contents were considered safe for adolescents. Drawing from the Japanese tradition, the Taiwanese *shôjo manga* artists have inclined to avoid graphic descriptions of heterosexual intimacy. The same concern appeared in Meng Chen's portrayal of Gertrude's sexuality.

In Meng's adaptation, Gertrude's sexuality is censored, and her outfit looks elegant and decently covers her body. In this interpretation, Gertrude is properly dressed even in her chamber when Hamlet interrogates his mother's chastity (III.4.20–24). There is no reference to Gertrude's sexual appeal, but an emphasis on her vulnerability (see figure 8.1). A close-up of Gertrude and Hamlet in panel 2 highlights that mother and son look alike. They both have long pointy nose and oval face; the mirror's reflection shown in panel 3 depicts Hamlet's cruelty to interrogate his mother's chastity. Hamlet cries out, "Disloyal Gertrude. Unchaste Gertrude. What is your relationship with my uncle?" in panel 4. These lines also bring to light the issues about Gertrude's moral defects and the conflict between mother and son.

In the West, sexual and violent implications in Shakespearean graphic novels of *Hamlet* are censored or edited out because their target readers are children and young adults. The western graphic novels deliberately remove Hamlet's obsession with Gertrude's sexuality to "protect their young minds from elements of immorality in the original text" (Rokison 66–67). Such a moral judgment also occurs in Meng Chen's adaptation, and she believes that Gertrude has no sense of guilt when she marries her brother-in-law, and the artist decides to make Gertrude feel ashamed after an intense argument with her son (Meng Interview).

In Meng Chen's adaptation, Gertrude dies after Hamlet accuses her of being an unfaithful wife and a shameless mother. Gertrude's punishment begins after she recognizes her guilt. She repents her doings and commits suicide (see figure 8.2). Her death, in the Christian viewpoint, cannot receive redemption because a person who commits suicide cannot be admitted to heaven. However, in Confucian culture, women must follow the "threefold dependence" (*Sancong*), depending on fathers, husbands, and sons at different stages of their lives. When a mother becomes a widow, she has to remain chaste and faithful to her deceased husband. In the Ming dynasty, women could receive a testimonial of merit for being a chaste widow, including those who committed suicide to avoid rape and forced marriage (Rosenlee 135). The widow's sexuality has been suppressed a long time under the influence of the Confucianism.

Unable to deal with the theme of incest, Meng Chen downplays Gertrude's sexuality and attempts to restore Gertrude's reputation as a virtuous woman. In her adaptation, Gertrude repents of her sins, and she climbs up to a tower, slips and falls off the tower. The scene looks like a suicide, which in the traditional Confucian perspective is a virtuous deed for a widow after her husband's death; she is worthy enough to be enshrined and worshipped. If a widow has to commit suicide to show her loyalty to her deceased husband, Gertrude's death is a testimony to her virtues.

On the other hand, a British *manga* version of *Hamlet* does not add a line and intends to offer an authentic storytelling as much as possible. The British *Manga Shakespeare: Hamlet* (2007) by Emma Vieceli is a perfect contrast for learners to compare how women's sexuality is portrayed in different cultures. Although *Manga Shakespeare: Hamlet* follows the *manga* style, it is not a typical *shôjo manga*. Vieceli's drawings try to capture violence, sexual tension and erotic implications in the original

Panel 2: "Look at this mirror in front of you. Who is she?" Panel 1: Pah!! "Ah!"
Panel 4: ". . . Use this mirror to see your conscience." Panel 3:
Panel 7: Pah!! Panel 6: Panel 5: "Hamlet, stop!" Pah!!
(Panels in Meng Chen's *shôjo manga* series of *Hamlet* are read from top down and right to left.)

Figure 8.1. Meng Chen's shôjo manga series of Hamlet (2006): Hamlet confronts his mother and questions her infidelity.

Meng Chen, Hamulete: Fuwangde Youhun (The King Father's Ghost). Taipei: Maxpower, 2006.

Panel 2: "I've failed as a mother." Panel 1: "The late King, our child, doesn't want me . . ."
Panel 3: "An unchaste wife who brings shame on you . . . How can I have the face to see you?"
Panel 8: "Your Majesty!" Panel 6: Panel 4:
Panel 7: Panel 5: ". . . You are? . . . Ah!"

Figure 8.2. Meng Chen's shôjo manga series of Hamlet (2006): Gertrude repents of her guilt after Hamlet scolds her for her infidelity.

Meng Chen, Hamulete: Fuwangde Youhun (The King Father's Ghost). Taipei: Maxpower, 2006.

play. Gertrude is not a cute, innocent girl, but a sophisticated woman with a healthy sexual desire. She does not fall into the *kawaii* category, cuteness in Japanese culture, but is the sexual object of the male gaze (Yiu and Chan 857). Gertrude's body is sexualized and fetishized into that of a young beautiful girl, *bishôjo*, in Japanese erotic manga. She is sexually appealing as the *bishôjo* who solicits sexual fantasies.

In *Manga Shakespeare: Hamlet*, Hamlet speaks explicitly to his mother, "Go not to mine uncle's bed" (III.4.175–176). In this *manga* interpretation, Hamlet holds Gertrude to the floor as if he intends to engage in sexual intercourse with his mother (see figure 8.3). Panel 1 gives a close-up of the mother and the son while panels 2, 4, and 5 give further details of their facial expression and panel 3 shows the same position at a different angle. The sexual tension between Hamlet and Gertrude, however, is not evident in Meng Chen's adaptation. The first-time learners may need guidance to read the original text and then to compare different interpretations from respective cultures. It is important to draw their attention to how Gertrude's sleeping with her brother-in-law is justified. In the original play, Gertrude is not given a chance to justify her remarriage, and, in this respect, she is "frail" in front of Hamlet's attack. She is punished by her conscience and then accidentally killed by drinking poisonous wine prepared for Hamlet. It is arguable whether Shakespeare intended to make her death portray moral justice. Consequently, the play allows scope for different interpretations.

In a comparison of the Taiwanese and British *manga* versions of *Hamlet*, the evidence of glocalization is quite visible. The Japanese coinage of glocalization claims that "the imperative for transnational companies is to 'Think locally, act globally,' that is, to distribute products globally and market them locally" (Iwabuchi 89). This glocalization strategy results in the successful reception of Japanese *manga* across the globe. A process of glocalization is demonstrated in the ways in which Taiwanese and British *manga* artists localize *manga* style into their specific cultures. As a result, the fashioning of Gertrude and her sexuality signifies the degree to which these cultures make use of the *manga* techniques and vocabulary to cope with their respective social attitudes toward the female body.

Traditionally, *manga*'s target audience is children and young adults; therefore, *manga* artists tend to omit the text that contains sexual and violent implications. This course gave a close study of the ways in which Gertrude's sexuality is filtered in these respective graphic novels for their target readers. Despite intercultural fusion in the *manga* style, these works demonstrate the social attitudes toward women's bodies and sexual desires in different cultures.

INTERCULTURAL SHAKESPEARE IN PERFORMANCE

The MOOC Global/Local Shakespeare emphasized that Shakespeare, originally from England, is now a British literature icon known globally. His works were translated, adapted, and appropriated in different languages on the stage all over the world. Via

Figure 8.3. Manga Shakespeare Hamlet (c) SelfMadeHero 2007. Illustrated by Emma Vieceli and Adapted by Richard Appignanesi.

Manga Shakepeare Hamlet © SelfMadeHero 2007. Illustrated by Emma Vieceli and Adapted by Richard Appignanesi.

MOOC, a diverse range of Western, Japanese, and Taiwanese productions are offered, in keeping with Shakespeare's "global" intentions. In *As You Like It*, Shakespeare compared the world to a stage and life to a play: "All the world's a stage, And all the men and women merely players" (II.7.142–143). The Elizabethan stage, therefore, represented the world. The Globe Theatre, built in 1599, was constructed to imitate a three-dimensional scale model of Earth, the world. In the Prologue of Shakespeare's *Henry V*, the Chorus invites the audience to imagine the theatre as the world:

> Or may we cram
> Within this wooden O the very casques
> That did affright the air at Agincourt?
> O pardon, since a crookèd figure may
> Attest in little place a million,
> And let us, ciphers to this great account,
> On your imaginary forces work.
>
> (Prologue 13–19)

The audience squeezed into this "wooden O," a quasi-O shaped theater. They used imagination to turn words into a great story. In fact, the Globe Theatre was not circular. It was an octagonal shape, built three floors high. Today, the Shakespeare's Globe, a replica of this Globe Theatre accommodates around 1,500 people.

This course began with an introduction to the 2012 World Shakespeare Festival, the biggest intercultural Shakespeare festival in history, held to enhance the international event of the Olympic Games. Thousands of artists from around the world presented nearly seventy productions, plus supporting events and exhibitions, right across the United Kingdom, and online. The Globe to Globe Festival, as part of the World Shakespeare Festival, hosted thirty-seven productions of Shakespeare's plays in thirty-seven languages at the Shakespeare's Globe. How did Shakespeare travel across borders and become so popular in the international community?

From a Taiwanese historical point of view, Shakespeare was first recorded and performed during the Japanese rule. However, local students seldom know this because a traditional Shakespeare course in Taiwan concentrated on the textual studies and rarely studied the texts in the intercultural context. This online course invited learners to look at how Shakespeare came to Taiwan in the timeline of world history.

Shakespeare came to Asia in the rise of Western Imperialism. His plays were frequently adapted in Japan in the late nineteenth century. It was also the time Japan transformed from an object of the Western Imperial gaze to a colonizer. In the early twentieth century, the Japanese playwright Emi Suiin adapted *Othello* into the Japanese context. The setting shifted from Venice and Cyprus to Japan and its new colony, the Penghu Islands (Pescadores) and Taiwan (Formosa). This localized production showed Japan's new identity as an emerging colonial empire and portrayed the newly established hierarchy in colonial Taiwan (Tierney 531). By performing this adaptation in Taiwan, the Japanese Empire was able to assert its particular political propaganda.

Because of political concerns, the Taiwan history had never been taught from a native inhabitant's perspective during the Japanese rule. The situation did not improve after the Chinese Nationalist Party (also known as the KMT[3]) represented the Alliance to take over Taiwan in 1945. The local history in relation to the Japanese rule was eliminated on purpose for many decades. Consequently, very few local Taiwanese scholars were aware that the first Shakespeare performance was in fact staged in the early twentieth century. To cultivate a learner's capacity of studying Shakespeare with intercultural awareness, this online course encouraged learners to discover the global and local dialectics of interpreting Shakespeare's texts.

Shakespeare's *Othello* tells a story of a Moorish general of Venice, Othello, who is tricked by Iago into believing that his fair Venetian wife, Desdemona, is having an affair with his lieutenant, Cassio. Othello's dark skin also evokes concerns and doubts because black represents evil.[4] In Emi's adaptation, the racial conflict in the original play is contextualized in the colonial hierarchy in Taiwan. The Moor General Othello becomes Muro Washiro, a Japanese colonial general of Taiwan. Washiro, who had formerly been an outcast,[5] becomes the most powerful person in the Japanese colony. In the last scene, before Othello commits suicide, he compares himself to "the base Indian" and to "a turbaned Turk" to emphasize his ignorance. In contrast, Washiro compares himself to a Taiwanese aborigine (*seiban*: literally, raw-savage), the lowest class in colonial Taiwan.

The Japanese adaptation of *Othello* represents an important testimony to the history of colonialism at the end of the eighteenth and the early nineteenth century. In the post-colonialist perspective, the global dissemination of Shakespeare is an outcome of the British cultural hegemony. The British Empire turned Shakespeare into an instrument of imperial authority to accomplish the process of cultural assimilation in its colonies. In order to assimilate the colonized to the colonizer's culture, the work of Shakespeare was regarded as the national that represented universal human values (Neill 168–169). Is Shakespeare universal? Taking the previous British colony of India, for instance, it is clear that the colonized's religious beliefs and ideologies are different from its colonizer's.

By studying the ways in which Shakespeare is localized and adapted into a specific locale or culture, the MOOC Global/Local Shakespeare learners became aware of how the local culture can take up the power of interpreting Shakespeare. In other words, they are able to read Shakespeare in a global context with an awareness of cultural differences. Consequently, this online course offered an introduction to Shakespearean adaptations across the world, including contemporary Shakespeare in Taiwan.

From the late twentieth century onwards, Shakespeare was adapted and appropriated on the Taiwanese stage. *The Kingdom of Desire*, based on Shakespeare's *Macbeth*, was performed in Beijing opera in 1986, the year before the lifting of Martial Law in Taiwan. After the success of *The Kingdom of Desire*, the Contemporary Legend presented *War and Eternity* in 1990 based on *Hamlet*. This adaptation was also performed in Beijing opera with a Chinese title called Wangzi Fuchou Ji, which literally means *The Revenge of the Prince*. The democratization in Taiwan also inspired local interpretations of Shakespeare.

Shamlet, performed in 1992, was the first parody of Shakespeare in Taiwan after the Japanese rule. This play was written, directed, and performed by Lee Kuo-shiu, a second-generation mainland Chinese. *Shamlet* was one of Lee's most successful productions, depicting a theater company's unexpected, disastrous mistakes while rehearsing and playing *Hamlet* on the stage. Shakespeare's *Hamlet* becomes a play-within-a-play in *Shamlet* and when the Fong-Ping theatre company (a parodic name of Lee's Ping-Fong Acting Troupe) rehearses and performs *Hamlet*, the play is frequently interrupted, reversed, and rewritten during performance. Actors shift from the persons putting on the play *Shamlet*, to their identities as actors in the story of the play, and to the Shakespearean characters in this failing production of *Hamlet*. The shuttling back and forth between different identities these actors play disturbs the plot progression. Furthermore, the ephemeral juxtaposition of several layers of identities they portray in the same time and space produces chaotic confusion and hilarity.

In the late twentieth century, the Godot Theatre in Taipei produced two Shakespearean musicals. Liang Zhimin, the director of the Godot Theatre, adapted *The Taming of the Shrew* into a live band musical *Kiss Me Nana* in 1997, and *A Midsummer Night's Dream* into *A Rock 'n' Roll Midsummer Night's Dream of the East* in 1999. Both musicals were written by well-known pop music composers, and performed by famous pop singers. When *Kiss Me Nana* was performed for the first time, people had to sing the national anthem of the Republic of China before a public event started. Godot Theatre played a rock-n-roll version of the national anthem for their rebellious youth, and the alternative version offended the conservatives.

In the twenty-first century, Shakespeare has been performed in various local languages and in many different regional opera styles. The Tainaner Ensemble performed two adaptations of *Macbeth* in Taiwanese, the most frequently spoken local language in Taiwan other than Mandarin Chinese. Both adaptations, *Sonata of the Witches—The Macbeth Verses* in 2004 and *Shakespeare Unplugged 3—Macbeth* in 2007, were based on the same script, which deleted the battle scenes and featured three witches as protagonists. The number of female characters thus increased, and an additional care was taken to show female influences. These Taiwanese adaptations created melody by emphasizing the rhyme, meter, and tone of the compositions. Actors were required to find rhythm in their body movements and the voices of their characters. The female actors playing Lady Macbeth and the three witches also incorporated different vocal techniques into their speeches to enhance the rhythm. The female voices came to the forefront and complemented the feminine power in these adaptations.

Golden Bough Theatre presented a kitsch adaptation of *Romeo and Juliet*, called *Yumei and Tenlai* in 2004. In search of an original Taiwanese theatre, *Yumei and Tenlai* imitated the popular "O Pei Ra style," a reformed Taiwanese opera performed during the late Japanese rule. The term "O Pei Ra" is derived from the Japanese pronunciation of opera. During the Sino-Japanese War, the traditional Taiwanese opera was banned. The Taiwanese opera performers had to wear the kimono and sing Japanese military songs to promote the imperial policies. The fusion of the Taiwanese opera tunes and modern western music became the hallmark of O Pei Ra, signifying a new theatrical practice in local popular culture. This O Pei Ra *Romeo and Juliet* recalled the memory of the good old days.

The aforementioned adaptations of Shakespeare are just a few examples to demonstrate the diversity of theatrical culture and languages in Taiwan. The universality of Shakespeare lies in the way he captures the whole possible range of human emotion. However, because of the differences in ideology, such as religious beliefs and customs, Shakespeare has to be adapted to fit well into different cultural contexts. Besides, Shakespeare is a British product that was often seen as being foisted on people in a colonial context. However, his materials were derived from many different cultures so people could recognize bits of themselves. To a certain degree, Shakespeare opened the world to his audience.

CONCLUSION

For over four centuries, Shakespeare's body of texts has been well received in theater, revered as a literary canon, and now popularized in comics and graphic novels across borders. Shakespeare's metamorphosis across cultural, linguistic, and geographical boundaries demonstrates an incredible journey of acculturation in this fast-growing and globalizing world. With the Internet, the MOOC Global/Local Shakespeare has encouraged online learners to develop intercultural competence through reading Shakespeare.

The MOOC Global/Local Shakespeare learners come from all over the world; therefore, the course served as a platform to improve one's cultural learning and cross-cultural communication. Furthermore, MOOCs provide in-video quizzes, a discussion bulletin board, and a virtual chat room to interact with learners. Compared with other distance-learning products, MOOCs are interactive because the in-video quizzes are autocorrected to give learners immediate feedback while the discussion bulletin board and virtual chat room offer both synchronized and asynchronized communication with instructors. In conclusion, this article offers many examples and new approaches to teaching Shakespeare in the twenty-first century.

Because the MOOC Global/Local Shakespeare was open to learners online, all learners presumably had Internet access by which they could search and watch the provided digital performance archives on websites. Motivated learners could initiate an online learning community in a chat room or a discussion forum on the MOOCs platform and share information and exchange ideas with other online learners across national borders. Learners were required to submit a case study of a non-English Shakespearean adaptation to explain how Shakespeare's play was adapted or rewritten to make it relevant to the target language and culture. This assignment was possible because several free online audiovisual archives, such as the Taiwan Shakespeare database, Asian Shakespeare Intercultural Archive, and MIT Global Shakespeares, were available online for free.

In this online course, learners developed and expanded their global vision by working and chatting with people from other cultural backgrounds. In the last class, they were required to submit a film of a soliloquy or a ten- to fifteen-minute scene and make the scene relevant to their own cultures. The short film was a creative assignment to integrate verbal and non-verbal communication skills by engaging linguistic, visual, auditory, gestural, and spatial semiotic systems that form a new model of text.

Today, learners living geographically apart are able to perform a scene online together if they wish. The online collaboration could be accomplished on Second Life (SL), a 3D virtual reality social networking website where users can build an ideal avatar in an online living environment. SL is free and users who download the SL viewer software are able to create their own avatars and settings. This virtual reality world enabled students to visualize, simulate, and role play a character. Online learners were able to collaborate with each other to perform a scene in the virtual world.

Like a typical MOOC course, MOOC Global/Local Shakespeare adopted peer assessment to grade student learning outcomes. A learner was given rubrics as the guidelines for accomplishing an assignment. The same rubrics were the criteria for assessment, helping a learner to act as a reviewer to assess a peer's assignment. The experience of being the reviewer helped the learners to critically reflect on each other's assignment. The dual role of being a reviewer and a reviewee motivated learners to develop responsibility for conscious learning.

Although MOOC Global/Local Shakespeare intended to create an interactive learning environment, it would not replace high-quality face-to-face instruction in the classroom setting. Due to network capacity and the time zone difference, the virtual chat room only allowed a limited number of learners to meet their instructors or peers online. This MOOC Global/Local Shakespeare course was incorporated into a for-credit Shakespeare course, taught at Providence University, Taiwan. The MOOC curriculum, including video lectures, assessments, and a discussion forum, were used to supplement the on-campus course. Students who signed up for this for-credit course needed to watch MOOCs online videos and attend project-based activities in the classroom. This blended learning model followed the emerging pedagogy, called flipped classroom, which reversed the traditional model of giving lectures in class and doing assignments at home.

In the MOOC Global/Local Shakespeare flipped classroom, students watched online lectures at home and then collaborated with their classmates to carry out research and complete different projects in the classroom. The local Taiwanese students who participated in this flipped classroom were the learners who benefited most because they were instructed to develop higher-order thinking skills by accomplishing project-based activities in the classroom. The flip teaching designed different activities to enhance student critical thinking, problem solving, team work, and creativity. Students had to complete collaborative or individual projects in class. The traditional lectures were replaced by teaching through questioning. These interactive activities enabled learners to work with their peers and reinforced their online learning.

Due to the pervasiveness of Internet and telecommunication, information occurs and spreads in various forms, such as images, charts, and graphics. However, despite the prevalence of images and visual media nowadays, it does not necessarily mean that "individuals are able to critically view, use, and produce visual content" (Introduction, ACRL Visual Literacy Competency Standards for Higher Education). The concept of literacy in the twenty-first century is no longer limited to the mastery of written and oral language. Visual literacy empowers individuals to participate fully in a visual culture. However, due to the emphasis of verbal literacy, there is no tradition to evaluate one's visual literacy in education.

The MOOC Global/Local Shakespeare flipped classroom encouraged students to improve both verbal and non-verbal communication skills. Visualization activities included students identifying and recognizing metaphors in Shakespeare's text, find appropriate visual images for the texts, and critiquing and evaluating choices made in the production of the images to construct meaning relevant to the metaphors. Through these activities, they were stimulated to develop critical-thinking and problem-solving skills.

In the digital age, MOOC Global/Local Shakespeare enables both native English speakers and non-native English speakers to study Shakespeare in an intercultural perspective through interactive learning designs. The frequent cultural exchanges brought about by the Internet broaches the issue of Shakespeare's universality while localizing Shakespeare becomes a global trend. These localized adaptations or appropriations not only make Shakespeare accessible for their cultures, but also enriched the ways Shakespeare is read and interpreted in this global village. By emphasizing a global and local perspective to examine Shakespeare's texts, this online Shakespeare course opens a new era of teaching and learning globally disseminated Shakespeare.

NOTES

1. All Chinese names in this paper are given in the Chinese order, which is family name goes before given name. Meng Chen's adaptation of *Hamlet* is published in two volumes, *Hamulete: Fuwangde Youhun* [*The King Father's Ghost*] and *Hamulete: Shiqude Daijia* [*The Price of Loss*]. Translations are all mine unless indicated otherwise.

2. Boys' love, often known as BL or yaoi, is a Japanese genre in *shôjo manga.*

3. The KMT refers to the Kuomintang, a political party founded by Sun Yat-sen shortly after the Xinhai Revolution of 1911 in China. The name literally means the Chinese National People's Party, but is more often known as the Chinese Nationalist Party.

4. In the Elizabethan period, blackness was often associated with evil and a visual signifier of eternal sin (Vaughan 54).

5. In the Meiji period, the feudal hierarchy was abolished and the lowest caste, Eta, which literally means the "filthy mass," becomes a "new commoner" (*Shin Heimin*), the newly liberated citizen from the lowest caste. Mura Washiro is the new commoner.

REFERENCES

ACRL Visual Literacy Competency Standards for Higher Education. Web. November 30, 2014, from http://www.ala.org/acrl/standards/visualliteracy.

Hayley, Emma. "Manga Shakespeare" in *Manga: Anthology of Global and Cultural Perspective*. Ed. Toni-Johnson-Woods. New York: Continuum, 2010. Print.

Iwabuchi, Koichi. *Recentering Globalization: Popular Culture and Japanese Transnationalism*. Durhan, NC: Duke University Press. 2002. Print.

Meng, Chen. *Hamulete: Fuwangde Youhun* [*The King Father's Ghost*]. Taipei: Maxpower, 2006. Print.

———. *Hamulete: Shiqude Daijia* [*The Price of Loss*]. Taipei: Maxpower, 2006. Print.

———. Interview by Yilin Chen. "10 Questions You Need to Know before Reading Meng Chen's Hamlet II." *MOOC Global/Local Shakespeare*. ShareCourse. Web. 2014.

Neill, Michael. "Post-colonial Shakespeare? Writing Away from the Centre" in *Post-colonialist Shakespeares*. Ed. Ania Loomba and Mark Orkin. London: Routledge. 1998. Print.

Rokison, Abrigail. "Shakespeare and the Graphic Novel," in *Shakespeare for Young People: Productions, Versions and Adaptations*. New York: Bloomsbury, 2013. Print.

Rosenlee, Li-Hsiang Lisa. *Confucianism and Women: A Philosophical Interpretation*. New York: SUNY Press, 2006. Print.

Shakespeare, William. *Henry IV, Part I*. New York: Macmillan, 1920. Print.

Shamoon, Deborah Michelle. *Passionate Friendship: The Aesthetics of Girls' Culture in Japan*. Honolulu: University of Hawaii Press, 2012. Print.

Tierney, Robert. "Othello in Tokyo: Performing Patriarchy, Race, and Empire in 1903 Japan." Shakespeare Quarterly 62.4 (2011): 514–540. *Project MUSE*. Web. 7 April 2015. https://muse.jhu.edu.

Vaughan, Virginia Mason. *Othello: A Contextual History*. Cambridge: Cambridge University Press. Print.

Vieceli, Emma, illus. *Manga Shakespeare: Hamlet*. England: SelfMadeHero, 2007. Print.

Yiu, Wai-hung, and Alex Ching-shing Chan. "'Kawaii' and 'Moe' – Gazes, Geeks (*otaku*) and Glocalization of Beautiful Girls (*bishôjo*) in Hong Kong Youth Culture." *Positions: East Asia Cultures Critique*. 21.4, 2013. 853–884. Print.

9

Modern Renaissance Education in Taiwan's Departments of Foreign Languages

Bernard Montoneri

I have been living and working in Taiwan for almost twenty years. When I was hired to teach at Providence University more than sixteen years ago, the department of English had six classes. I was reminded of this fact last year when one of my former students from Class F came to visit me in my office in the need of a letter of recommendation. For various reasons, such as the *titanesque* (that is both gigantic and tragic) decline of the birth rate in Taiwan (the lowest in the world in 2010: 0.83; only 0.84 in March 2012; Montoneri et al. 2012; Wang & Basten, 2013), we have seen our department shrinking from six to three classes over the years. Sadly, the administrators have decided to cut one more class, meaning that the Department of English will have only two classes left by 2016. It was also decided to close some classes of literature: for example, one section of British Literature and one section of American Literature have already been cancelled. This article is more of a discussion that offers to reflect on how to preserve the study of literature, not only in the Department of English at Providence University, but in other departments of Foreign Languages facing the same challenges in Taiwan and abroad. It seems that if countries in Asia wish to grow and blossom, they should analyze and understand how Europe, after more than a century of wars (The Hundred Years' War, 1337–1453) and disease (the Plague), managed to become the economic, political, and cultural center of the world during the Renaissance. How could Taiwan and notably Taiwanese students benefit from emulating some of the values and ideals that made Europe so successful? How could we incorporate some of the virtues of Renaissance people into our curriculum? What if our students, at least some of them, could become Renaissance men and women, and what if they were exactly the type of graduate students CEOs wished to hire? Finally, how can we adapt and update Renaissance virtues to the twenty-first century by using information and communications technology (ICT) and notably social and professional networks such as Facebook and LinkedIn?

HISTORICAL BACKGROUND

The Middle Ages officially begin with the collapse of the Western Roman Empire (Rome, AD 476) and end with the collapse of the Eastern Roman Empire (Constantinople, 1453, also the date of the end of The Hundred Years' War). Despite all the wars, the crusades and the inquisition, the medieval period is also marked by the founding of universities; the term university is derived from the Latin *universitas magistrorum et scholarium* ("community of teachers and scholars"). Among the most ancient universities in Europe are the University of Bologna (1088), the University of Paris (c. 1150), the University of Oxford (1167), the University of Cambridge (1209), and the University of Montpellier (1220). The University of Bologna began as a law school teaching Roman law. The beginnings of the Renaissance are marked by the interest in the potential knowledge that could be gained from ancient Greek and Roman texts. The term Renaissance is a French word meaning rebirth or reborn; the Italian equivalent is *Rinascimento*, from *rinascere* ("to be reborn"). What is reborn is the spirit of Ancient Greece and Rome. *Reborn* also refers to the baby boom that Europe experienced at the end of The Hundred Years' War. The population was young, dynamic, curious, and eager to learn and to discover. As a result, the Renaissance is a rich and complex period that blends various movements such as the Age of Discovery (Columbus, 1492), a baby boom, the birth of Protestantism, the propagation of universities and of course humanism. The emergence of humanism, a philosophical and ethical stance but also a method of learning, was linked to the idea that society, and notably the rulers, could benefit from the expertise generated from these educational institutions. The humanists believed in the education of a universal man or *uomo universale*, an ancient Greco-Roman ideal of an honorable person combining intellectual and physical excellence (*Mens sana in corpore sano*: "A sound mind in a sound body"; *Satire X* of the Roman poet Juvenal; Hendry 2005). In order to achieve this goal, the education during the Renaissance emphasized ancient literature and history because classics were believed to notably teach moral instruction. Humanists were trained to speak and write with eloquence and clarity. The *studia humanitatis* or "study of humanities" included Greek, moral philosophy, history, and literature (notably poetry). The fourteenth century saw the emergence of the emancipation of the individual. According to Kreis (2009), "The writings of Dante, and particularly the doctrines of Petrarch and humanists like Machiavelli, emphasized the virtues of intellectual freedom and individual expression." It was believed during the Middle Ages that human beings, because of their sinful nature, could not attain excellence (Augustine's *City of God*, 426 AD). Humanism can be considered as an educational and cultural program putting emphasis on individual achievement through hard work and on the creative powers of mankind.

Baldassare Castiglione (*The Book of the Courtier*, 1528) used the term *sprezzatura* ("an easy facility in accomplishing difficult actions which hides the conscious effort that went into them" in Rebhorn 1978, 33) to describe the correct attitude that should accompany the various talents of a polymath: having a vast knowledge of the humanities, Classics and fine arts; having a warrior spirit; being eloquent and calm; giving an impression of effortlessness and ease even though it is the result of very hard work (Saint-Exupéry excelled at this in writing *The Little Prince*).

THE BETTERMENT OF MAN AND THE PRACTICE OF VIRTUES

"The one true path that leads to a tranquil life is that of virtue" claims Juvenal in the same *Satire X* quoted above (Kline, 2011). In Ancient Rome, the term *virtus*[1] first meant "manliness" (from *vir*, "man"). It is related to masculine strengths of excellence, courage, honor, character, and worth. It also includes prudence, justice, and temperance. During the first semester of the academic year 2014–2015, I introduced *The Republic-Book VI: The Philosophy of Government* by Plato (390 BC). Concretely speaking, practicing virtues begins with reading, analyzing, and connecting to our times various virtues defined by the classics. Plato proposes to define the nature of the philosopher. During the class we discuss the various qualities and virtues listed in the extract, and I help the students to observe that the word *love* is repeated many times: "love knowledge," "lovers of all true being," "love the truth," "lover of wisdom," and so on. The philosopher is able to discern absolute goodness and truth.

In Rome, Cicero (106–43 BC) proposed a theory of virtue (the 4 cardinal virtues), which can be found in *De re publica* ("On The Republic") and in *De Inventione* ("On Rhetorical Invention"): "For virtue is a habit of the mind, consistent with nature, and moderation, and reason. Wherefore, when we have become acquainted with all its divisions, it will be proper to consider the whole force of simple honesty. It has then four divisions—prudence, justice, fortitude, and temperance." (Cicero 1888. II, LIII. English translation) According to Baldwin (1990): "Cicero's Stoic humanism offered a more appealing model for Renaissance humanists eager to fuse political and moral virtue and to defend the moderate enjoyment of earthly pleasures."

Interestingly, the cardinal virtues (Rickaby, 1913) are also listed in the Old Testament. The deuterocanonical book *Wisdom of Solomon* notes, "She [Wisdom] teaches temperance, and prudence, and justice, and fortitude, which are such things as men can have nothing more profitable in life." (*Wisdom of Solomon* 8: 7; King James Bible online 2014).

Not surprisingly either, in many versions of the tale of King Arthur including Sir Thomas Malory's *Le Morte d'Arthur* of 1485, the ideal of the virtuous king refers to the importance of truth and justice. In *Excalibur* (1981), the best cinematographic adaptation of Malory's novel, Merlin famously answers Arthur's question:

> Arthur: "Which is the greatest quality of knighthood? Courage? Compassion? Loyalty? Humility? What do you say, Merlin?" [. . .]
>
> Merlin: "All right, then. Truth. That's it, yes, it must be truth, above all. When a man lies he murders some part of the world. You should know that."

I made a Chinese translation of the dialogue and posted it on my website.[2] I notably used it in 2008 during the English Festival at Providence University for the translation contest. After reading this extract, we would naturally talk about the issue of cheating. With irony, *Business Insider* notes that some students in Mainland China are developing new skills, that is using technology to build cheating devices for the exams. I wonder how Taiwanese students would react to the infamous slogan chanted recently by some students in Mainland China: "We want fairness. There is no fairness if you do not let us cheat."[3] I also wonder whether CEOs would want to hire this type of student.

To become responsible and virtuous adults, students need to learn about great thinkers and artists who are set as examples to emulate. After all, fame, honor and respect are the privilege of honest and hardworking people.[4] In September 2014, I began my American Literature class by introducing *The Autobiography of Benjamin Franklin* and his famous list of thirteen virtues.[5] Franklin suggested providing a model of self-betterment written while he was living in Paris in the 1780s. Interestingly, many of these virtues have a Roman flavor, such as temperance, justice, and cleanliness. During the Renaissance, Roman virtues were redefined to focus on active, virile and civic capacity: freedom-loving, spirited and virtuous Italian and Florentine people should live in the only suitable form of government: the Republic. Franklin had surely been greatly influenced by the Ancients (Plato and Cicero) and by the ideals of the Renaissance. But the Founding Father quickly acknowledged with humor he could not perfect these thirteen virtues ("A benevolent man should allow a few faults in himself, to keep his friends in countenance" (Franklin, 1771–1790); however, his endeavor made him much happier. He thus wished his descendants to feel the same enjoyment and benefits from practicing these virtues.

Beyond a Good Learning Attitude: Educating Renaissance Men and Women

According to Rakowski (2011), "High performance in school should be focused on discovering and developing true polymaths—students of multiple talents, skills, and experiences, as well as a strong personality." The Renaissance polymath (Greek, "having learned much") is a person of many skills who possesses an artistic and creative spirit, as well as a vast knowledge, notably in history, geography, literature, and in the arts, an unquenchable curiosity and a wonderful creativity. Versatility, curiosity, feeling of wonder, spirit of adventure, qualities we have to ignite in our student's mind. I believe courses of literature can do this. The Ministry of Education in Taiwan and universities also have a fundamental role to play in order to help students, to travel abroad for example (grants, exchange programs).

Obviously, few people in the world would dream to be Renaissance polymaths like Leonardo Da Vinci (1452–1519), who was a painter, sculptor, architect, musician, mathematician, engineer, inventor, anatomist, geologist, cartographer, botanist, and writer. Well, it is overwhelming. However, we can teach students to understand, appreciate, and emulate Renaissance qualities and values such as exploring the world and traveling, being an erudite, developing artistic traits and imagination, being a volunteer, and improving communication (eloquence) and writing skills. Renaissance and humanist artists and thinkers were flexible, adaptable, and versatile. Qualities, as we shall see, which are in high demand on the job market. But to become a Renaissance man or woman means years of hard work and dedication at school.

Peterson and Seligman (2004) offer a classification of the psychological traits of human beings notably based on the four cardinal virtues and on Christian values, including *wisdom and knowledge* (creativity, curiosity, open-mindedness, love of learning, perspective and wisdom), *courage* (bravery, persistence, integrity, vitality), humanity (love, kindness, social intelligence), *justice* (active citizenship / social responsibility / loyalty / teamwork, fairness, leadership), *temperance* (forgiveness and mercy, humility and modesty, prudence, self-regulation and self control), and *transcendence* (appreciation of beauty and excellence, gratitude, hope, humor and playfulness, spirituality).

Interestingly, creativity is personified by Leonardo da Vinci and Albert Einstein and love of learning by Benjamin Franklin. Social intelligence, a self- and social-awareness, consists in befriending others, a trait anyone nowadays can practice in social and professional networks.

Rakowski (2011) rightly says that universities, administrators, and educators should propose a new philosophy of education. She suggests to focus on ten areas to educate Renaissance polymaths who will become erudites, entrepreneurs, soldiers, teachers, writers, artists, linguists, computerists, volunteers, and handypersons.

My suggestion to educate polymaths would be to use literature to teach *moral and intellectual virtues*. We should personify these virtues and introduce books focusing on them, as follows:

Human and moral virtues

a) Love: including compassion, gratitude, humility, respect, and forgiveness. Examples: *Les Misérables* and *Notre Dame de Paris* by Victor Hugo; *The Count of Monte Cristo* by Alexandre Dumas (revenge versus forgiveness); *The Iliad* by Homer (Book 24 Priam Ransoms Hector's Body).
b) Courage: including a fighting spirit (not giving up, persistence) bravery, and vitality. Fairy tales offer numerous examples of young men and women facing adversity who refuse to give up and who are rewarded for their courage. I notably enjoy introducing *The Wizard of Oz* and the wonderful spirit of Dorothy.
c) Truth: including honesty, confidence, critical thinking, justice and freedom, respect for history and original sources. Truth can be connected to *Beauty and the Beast* ("That is true," said Beauty, "for I cannot tell a lie") and the famous magic mirror that never lies in *Snow White*. Knights of the Round Table and even Superman fight for *Truth and Justice*.
d) Curiosity: including love of learning, versatility, creativity, self-motivation, and spirit of adventure, which are valuable Renaissance virtues and qualities. In European Literature, we study *The Odyssey* by Homer and *The Travels* by Marco Polo.

EASTERN AND CHINESE VIRTUES

Educators must ensure that Asian and notably Taiwanese students feel that learning Western literature is valuable and can help them in their everyday life. It is important to relate stories studied during the class to their personal experience and to their own culture. They will quickly realize that Western and Eastern values and virtues are very close and that the past can teach them useful advice, applicable and concrete values, and skills. For example:

1. Buddhism as well as Christianity teaches about generosity (as opposed to greed). When we study *A Christmas Carol* (Charles Dickens, 1843), I show students that the Christian virtue of generosity is very similar to the one introduced in the *Diamond Sutra* (金剛經 *Jingang Jing*). The *Dana paramita*, or perfection of giving, is the first of the six *paramitas* ("perfections" or "virtues"), that bodhisattvas ("enlightened beings") practice. The other virtues are morality, patience, energy, concentration, and wisdom. The antidote to greed is material generosity, food, and medicine. In *A Christmas Carol*, we can see the transformation of Mr. Scrooge from a lonely and greedy man to a compassionate and generous family man who cares about poor boy Tiny Tim in need of food and medicine to survive.
2. One of the six virtues in the *Diamond Sutra* is also very close to Western virtues: Śīla or "honesty" (related to truth). Many fairy tales and romances of the Middle Ages (*Beauty and the Beast*, *King Arthur*) discuss the notion of truth and honesty. Moreover, many Roman virtues, as well as Franklin's virtues, correspond to what the Buddhists call Śīla: moral conduct and character, code of ethics, moral principles and precepts, discipline, and self-restraint. According to Buddhism, the reward and blessing of morality is freedom from remorse, tranquility, vision, and knowledge with regard to Deliverance.
3. *Prajñā* or "wisdom" is the ultimate virtue in Buddhism, notably in the *Mahāyāna* ("Great Vehicle") branch. To attain wisdom, Buddhist practitioners as well as Greek philosophers are to cultivate various virtues first, such a compassion, generosity, moral virtue, and love of knowledge and truth. We discussed earlier in this article how Plato proposed to define the nature of the philosopher and how he insisted on the virtue of love. This notion of wisdom could also be compared to other essays and novels. The most famous examples of wisdom literature are to be found in the Bible, notably in the *Book of Job*, the *Psalms*, and the *Book of Wisdom*. Children's books in the West, such as Saint-Exupéry's *The Little Prince*, are also rich in insight and wisdom.
4. Confucianism and filial piety. Some of the most important themes in Eastern and Western literature are love of family, filial piety, and the importance of the father figure. The character *xiao* 孝 (piety) represents an elder (老 *lao*) being carried by a son (子 *zi*). The *Book on filial piety* (*Xiaojing* 孝經) is a Confucian book notably focusing on the relationship between father and

son. It was probably compiled by disciples of Confucius or by Confucian scholars during the early Han period 漢 (206 BCE–220 CE). Filial piety is the foundation of all other virtues. The *Twenty-four Filial Exemplars* (*Ershi-si xiao* 二十四孝) depicts how children should exercise piety. Most stories in Western literature begin with a tragedy: the father is either weak, sick or dead. Children Literature for example teaches children to consider tragic events as an opportunity to learn, grow, and succeed (*Cinderella*, *Puss in Boots*). In Collodi's story, if Pinocchio, the lazy troublemaker, self-centered and distractible puppet is rewarded by the Blue Fairy and becomes a real boy, it is only because he sacrificed his life to save his father Geppetto. Children are supposed to save a psychologically or physically weakened father or to carry on his legacy and restore his honor if he dies. Many classical stories in the West, such as *The Aeneid*, *The Iliad*, and *The Odyssey*, are also close to the virtue of *xiao*: the Latin word *pietas*, meaning "duty," "loyalty," and "filial piety," was one of the main virtues in Ancient Rome. It was notably the virtue of the founding hero Aeneas in Vergil's epic. Another famous example of filial piety can be found in *The Iliad* with Priam begging Achilles to return the body of Hector. King Priam, Hector's father and the ruler of Troy, ventures into the Achaean camp and tearfully pleads with Achilles to take pity and return Hector's body ("Think on your own father and have compassion upon me, who am the more pitiable, for I have steeled myself as no man yet has ever steeled himself before me, and have raised to my lips the hand of him who slew my son"; *The Iliad*, book 24). After Priam invokes the memory of Achilles's father, Peleus, Achilles returns Hector's corpse to the Trojans. Priam and Achilles both weep for their loss. According to Shankman and Durrant, "It would be wrong to say that we do not find this trait of *xiao* exhibited in Homer's *Odyssey*. The first four books of the poem—the Telemachy—amply record Telemachus's desire to find and even emulate his father."

5. Taoism and Taoist virtues. The Three Treasures (*sanbao* 三寶) are basic virtues in Taoism. In the *Tao Te Ching* 道德经 (written around sixth century BC by Laozi 老子), they are compassion (and kindness, love *ci* 慈), frugality (moderation, simplicity), and humility. *Tao* (or *Dao*) is the "way" and *Te* (or *De*) is 'virtue." *Daode* 道德 literally means "ethics," "ethical principles," "morals," or "morality." Interestingly, the *Tao Te Ching* parallels *ci* (translated then as "parental love") with Confucian virtue *xiao* (孝).

A NEW RENAISSANCE IN LEARNING AND TEACHING TWENTY-FIRST CENTURY CLASSROOM ETIQUETTE

It is important when practicing these virtues and skills to remember that Franklin said with humor we cannot achieve perfection and should remain modest and open-minded. We should not give educators more pressure than they already have, and they should not be over demanding. We should also adapt these traits to the problems, challenges, and tools of the twenty-first century.

No doubt that educators have always complained about their students' behavior and ignorance. In order to build a harmonious teaching and learning environment, students should follow a classroom etiquette, such as not being late, attending class regularly, not eating and drinking during the class, not sleeping, preparing for the class and participating, showing respect to the teacher and to the other students. But in the twenty-first century, we face new challenges: with the advent of social networking and the smartphone, the situation is more complicated. The good (or bad?) news is that students do not sleep anymore during the class: they use their cellphone. And they use it everywhere, all the time, while walking in the streets, while eating with their friends, during the class. There is already a lot of research on health-related side effects of cellphones: regular use notably causes sleep disturbances and symptoms of depression (Yen et al. 2009; Takao et al. 2009; Thomée et al. 2011). There is even a new disease named after this constant online activity ("nomophobia," short for "no-mobile-phone phobia") and after the constant uploading of selfies (Body Dysmorphic Disorder). A recent study shows that cellphone addiction is also posing potential risks for academic performance among college students (Roberts et al. 2014). Interestingly, in the Henan province of China, students at Luoyang Institute of Science and Technology chose to voluntarily turn off their phones and to put them into wall pockets with name tags before classes (Chinanews 2014).

A New Renaissance in Research

At the end of 2012, I was invited by the department of foreign languages at Da-Yeh University in Changhua County, Taiwan, to give a lecture on how Renaissance values and virtues might be beneficial to students and teachers on the island. This event, which clearly shows the growing interest in the notion of New Renaissance in Taiwan, inspired me to write this article. During the presentation, all the teachers seemed to agree that students and educators alike need to be more open-minded and eager to become knowledgeable in more than just one field. When possible, students, for example, should be double majors and have a good command of at least two foreign languages.

As for scholars, they would without doubt benefit from cooperating with colleagues who have complementary skills and expertise. With humility, we must acknowledge our ignorance in some domains. However, it is not a reason to give up and to focus on the study of just one field. For example, I did not study statistics at school; as I wanted to do research in the field of education, I proposed to other scholars to build a research team and to publish original studies using Data Envelopment Analysis (DEA); each one of us brought his or her expertise and experience. Educators too should rethink the way they do research. There are so many scholars, so many teams, and so many journals nowadays around the world. To be competitive and to publish original research with a clear contribution, we need to work together, do cross-field research, and if possible cross-culture and cross-language.

ADAPTING RENAISSANCE VIRTUES TO THE TWENTY-FIRST CENTURY

Why should we go beyond adopting some basics rules about classroom etiquette? What are the benefits of practicing Renaissance virtues and how can we adapt them to this new century? A good etiquette may help young adults study in a positive learning environment and make the most out of their education (which is, in Taiwan private universities, very expensive).

These qualities should be adapted to the twenty-first century by including advanced computer proficiency and knowledge of the modern job market by using Facebook or LinkedIn, by applying to programs and graduate schools abroad, and by promoting exchanges with foreign students and professors. We also need to focus on *practical skills* that are essential whatever profession students decide to exercise. Some practical skills in high demand on the job market:

1. Teamwork and problem-solving ability are often listed by CEOs in various countries as the most important employment skills. Reading classic dialogues written by thinkers and philosophers may help develop these skills (Plato, Aristotle).
2. Communication skill includes presentation skill, eloquence, negotiation, social and business etiquette, cultural, social, and gender differences. British Literature here will obviously be very useful (think about BBC TV series: *Pride and Prejudice*, *Jane Eyre*, *Wuthering Heights*, and so on).
3. Computer skill, social and professional networking. Rakowski (2011) calls it "computer savvy: advanced computer proficiency." Computer skill and networking are a must nowadays, notably when looking for a job. Obviously most students nowadays use a computer, a cellphone, and spend a lot of time online. However, it does not really count as being computer savvy: Rakowski considers it combines the knowledge of computer hardware and of up-to-date software applications.

Roger Lee, a Chinese from Beijing working in Taipei, posted on LinkedIn an article in which he asks the following question: "What do we need in the real world?" He proposes to focus on five skills to help students "survive" in the business world: Leadership Skills / Work with teams; Communication Skills / Command of Language; Basic Math and Logical mindset; Resourcefulness and Independent; and Ability to Unlearn and Relearn (Lee 2014). This corresponds quite well to the virtues and skills we have discussed so far. Students need a vast knowledge that can be unlearned if necessary (being tolerant and open-minded, logical, practicing critical thinking), they should be versatile, sociable, and possess the most valued skills on the job market: teamwork and communication skill. Many surveys and studies have been published to emphasize the importance of these skills.

In the UK, the promotion of students' employability skills is considered a priority for higher education (Little 2006). In a recent survey, about 54 percent of business leaders in Europe think young people lack soft skills such as confidence, teamwork, self-motivation, networking, and presentation skills (FreshMinds Research 2011). In Taiwan, 74.4 percent of the respondents in a local study believed that lack of proper attitude and employability skills are the main reasons for highly educated jobless situation (Yan 2007). A study conducted by Hsieh and Ho (2009) based on a survey of 420 Taiwanese employers indicated that employability factors such as personality, problem solving skills, and insight are among the first most important required qualities for employment.

In April 2013, the Association of American Colleges and Universities (AAC&U) released a survey of employers and promoted a new compact between college presidents and employers in favor of a broad, liberal-arts education for all. The document shows that "93 percent of employers say that a demonstrated capacity to think critically, communicate clearly, and solve complex problems is more important than a candidate's undergraduate major." According to the compact, "80 percent of employers agree that, regardless of their major, all college students should acquire broad knowledge in the liberal arts and sciences" (LEAP Employer-Educator Compact 2013). As of March 2014, 100 college presidents and 170 business and nonprofit leaders had signed on to the LEAP Employer-Educator Compact. "A successful liberal arts education develops the capacity for innovation and for judgment," said Michael Roth, President at Wesleyan University, on the blog of *The Huffington Post* (Roth 2009).

BUILDING CONNECTIONS AND APPLYING FOR A JOB

During the second semester of 2013, the president of ESHotel, Arnaud Bouvier, came to Taiwan and notably to Providence University to meet with students and hopefully to recruit some of them: ESHotel is a school in Paris and London that offers international hotel and tourism management programs. I attended his presentation and as usual I invited him to connect with me on LinkedIn. Arnaud Bouvier's specialties, as listed on his LinkedIn profile, are problem solving and management training. When he talked to our students, he said that actually his school is more interested in recruiting undergraduates from departments of Foreign languages or from departments of English than from management departments. According to him, they make better managers because they are trained to exercise the most valuable skills on the job market (and to run a school), that is communication skill and relationship management skill. In fact, most companies need graduates with a good attitude and a good command of English (vital in the field of tourism).

Using LinkedIn is becoming a must and it is very surprising that few students in Taiwan know about it. The world's largest professional network (313 million members in over 200 countries and territories; LinkedIn Press Center 2014) might help students connect, that is to create the positive spin of the Asian concept of *guanxi* ("connections").

A Renaissance education also means benefiting from modern technology and the fast growth of social and professional networking. For example, LinkedIn grew from 32 million in 2009 to over 300 million worldwide by mid-2014 (Pauley 2014). The widening communication circle in Foreign Language education might help students connect to each other, to teachers, and to CEOs from almost every country on Earth. LinkedIn has over the years become a platform to inform, influence, and educate. There are over 39 million students and recent college graduates on LinkedIn (LinkedIn Press Center 2014). As a matter of fact, students, teachers, and educational institutions can nowadays benefit from using LinkedIn by sharing posts and publications, joining and creating groups (notably language groups, local and global group, groups offering jobs and advice to apply for a job), building research teams, asking for a letter of recommendation, following institutions, recruiting students, finding an adviser or a mentor.

Most CEOs nowadays acknowledge that they are using LinkedIn to hire graduates (Martin 2014). According to Forbes (Adams 2013), "while 90 percent of students use Facebook and 78 percent are regularly on YouTube, nearly half—46 percent—say they never use LinkedIn." Dan Schawbel, managing partner of Millennial Branding, states that "it's an absolute must to create a LinkedIn profile by the time you're a senior in college, but it's an even better idea to start as a freshman." (Adams 2013).

CONCLUSION

A quite surprisingly large number of administrators (and in my experience a refreshingly small number of students) think that the study of literature is irrelevant on the job market in our twenty-first century. More and more departments of languages use the word *applied*, an obvious attempt to declare that literature is not. To convince my students and hopefully some education administrators, I would say the following:

1. The list of the 100 most influential people in the history of the world, even though controversial (too many Westerners, few women), is interesting because of its inclusion of a large number of writers and thinkers (Sterbenz 2013).[5] Their writing changed the world, caused revolutions, and were often banned (the publication of *Index Librorum Prohibitorum* or *Index of Forbidden Books* was officially suppressed in 1966). Another alternative list (Pettinger 2013) includes, for example, a larger number of women activists, leaders, scientists, and artists (but almost no female writers: Mother Theresa, Rosa Parks, Queen Elizabeth II, Marie Curie, Queen Victoria, Cleopatra, etc.). Here again, very few Asian people

are included; no mention of Confucius, Laozi, or other important figures in the East. However, because China is quickly becoming a world power, people will surely pay more attention to Asian thinkers, writers, and artists.

2. The study of Western Literature and philosophy in Taiwan's Departments of Foreign Language can teach students around the world about common values and virtues such as compassion, freedom, truth, equality, and justice. Students need to be given the opportunity to discuss freely and to practice critical thinking. They need to learn problem-solving skills and communication skills; they need to enrich their vocabulary to be able to express their opinion and feelings. Moreover, in Taiwan, while teaching Western Literature, it is particularly interesting and enriching to compare Western and Eastern values and virtues. they might realize that, actually, there is little difference. We all long for love, friendship, family, freedom, compassion. They might also see that feelings and important matters have not changed much over the past thousand years (filial piety, for example, as expressed by Confucius or in *The Iliad*).
3. Most movies (*The Shawshank Redemption*, *The Lord of the Rings*), operas (*Carmen*, *Rigoletto*), paintings (*The Bible*, *The Iliad*), and musicals (*The Phantom of the Opera*, *Les Misérables*) are adapted from novels. Because of the American historical fantasy drama series "Da Vinci's Demons" (Starz and BBC Worldwide), Leonardo da Vinci is more famous than ever. The series presents a fictional account of his early life and portrays *l'artista* ("the artist": a person whose creative work shows sensitivity and imagination) as an energetic Renaissance man already renowned as a genius but yearning for even deeper knowledge.
4. Students need a liberal education and the practice of Renaissance virtues because it is exactly what most CEOs around the world want. Above all, graduates are hired for their personality and attitude, not for their scores during sophomore year.

I will conclude with a quote. Here is what Richard Branson (founder at Virgin Group) said in 2013; I italicized keywords:

> You can learn most jobs extremely quickly once you are thrown in the deep end. Within three months you can usually know the ins and outs of a role. If you are satisfied with the *personality*, then look at experience and expertise. Find people with *transferable skills*—you need *team players* who can pitch in and try their hand at all sorts of different jobs. While specialists are sometimes necessary, *versatility* should not be underestimated.

NOTES

1. The motto of Providence University is *Virtus cum Scientia* (Latin for "virtue and knowledge"). The words were chosen by Saint Mother Théodore Guérin (1798–1856), foundress of the Sisters of Providence of Saint Mary-of-the-Woods, Indiana. Anne-Thérèse Guérin, born

in Brittany, France, was beatified in 1998 by Pope John Paul II and canonized a saint in 2006 by Pope Benedict XVI.

2. Extract of *Excalibur* (1981). Interior Hall of the Round Table, Camelot. The movie script can be viewed online: http://www.imsdb.com/scripts/Excalibur.html.

3. "Chinese Teens Have Found Remarkable High-Tech Ways to Cheat on Tests." Please see https://news.vice.com/article/high-stakes-testing-led-chinese-teens-to-use-spy-like-cheating-equipment#ixzz3BZFhZfN5.

4. Please see "Even Geniuses Work Hard" by Carol S. Dweck. This psychology professor at Stanford University states that people with a *fixed mindset* believe they are born intelligent and don't expect any growth. People with a *growth mindset* think that they can develop their intelligence and consider challenging work as a chance to learn and to grow. She suggests that teachers should create a *growth-mindset culture* in the classroom. Her learning theory was endorsed by Bill Gates on August 26, 2014. Please check http://www.businessinsider.com/carol-dwecks-growth-mindset-theory-tweeted-by-bill-gates-2014-8 and http://www.mrscullen.com/images/wd_hw_evengenuisesworkshard.pdf.

5. The thirteen virtues are temperance, silence, order, resolution, frugality, industry, sincerity, justice, moderation, cleanliness, tranquility, chastity, and humility. Please see http://mengwong.com/franklin13-3.pdf.

REFERENCES

Adams, Suzanne. "5 Mistakes College Job Seekers Make." *Forbes*. 22 April 2013. Web. 9 September 2015.

Baldwin, Robert. "Humanism and the Early Italian Renaissance." New London: Connecticut College. Essay written 1990 and revised annually. Web. 9 September 2015.

Bible, King James. "Wisdom of Solomon," from *The Holy Bible*, King James version (Apocrypha) Electronic Text Center, University of Virginia Library. Web. 10 October 2015.

Branson, Richard. "How I Hire: Focus On Personality." LinkedIn. 23 September 2013. Web. 10 October 2015.

Chinanews. "Smartphones in Class? Let Students Decide." 4 April 2014. Web. 10 October 2015.

Cicero, Marcus Tullius. *The Orations*. Literally translated by C. D. Yonge. London: George Bell & Sons. Vol 4: 241–380. 1888. Web. 10 October 2015.

Franklin, Benjamin. *The Autobiography of Benjamin Franklin*. Independence Hall Association, Philadelphia, Pennsylvania. 1771–1790. Web. 10 October 2015.

FreshMinds Research. "Closing the Gap between Business and Education. Survey report." London: FreshMinds Limited. 2011. Web. 10 October 2015.

Hendry, Michael. "On-line Text of Juvenal's Satires. Satire X." 4 October 2013. Web. 20 October 2015.

Hsieh, H. C., & Ho, L. K. "Employee Competencies of Taiwanese International Trade." *Commerce and Management Quarterly* 10.4 (2009): 707–733.

Kline, A. S. "Juvenal, the Satires. Satire X— The Vanity of Human Wishes." 18 February 2011. Web. 20 October 2015.

Kreis, Steven. "Renaissance Humanism." *The History Guide*. 3 August 2009. Web. 10 October 2015.

LEAP Employer-Educator Compact. "Making Quality a Priority as Americans Go to College." April 2013. Web. 10 October 2015.

Lee, Roger. "Are Schools and Parents Killing the Job Market ?" 2 September 2014. Web. 10 October 2015.

LinkedIn Press Center. "About LinkedIn." Newsroom 2014. Web. 10 October 2015.

Little, B., & ESECT. *Employability and Work-based Learning*. York, United Kingdom: The Higher Education Academy. 2006. Print.

Martin, Emmie. "Here's Why This CEO Finds All Her Employees through LinkedIn." *Business Insider*. 27 August 2014. Web. 10 October 2015.

Ministry of Education (MOE). "Education Statistical Indicators." Republic of China, Taiwan. 2015. Web. 25 October 2015.

Montoneri, B., Lin, T. T., Lee, C. C., & Huang, S. L. "Application of data envelopment analysis on the indicators contributing to learning and teaching performance." *Teaching and Teacher Education* 28.3 (2012): 382–395.

Pauley, Ron. "LinkedIn for Academics." 9 September 2014. Web. 25 October 2015.

Peterson, Christopher & Seligman, Martin E. P. *Character Strengths and Virtues: A Handbook and Classification*. Oxford: Oxford University Press. 2004. Print.

Pettinger, Tejvan. "List of Top 100 Famous People." Biography Online. 2 January 2013. Web. 10 October 2015.

Rakowski, Joanna. "The New Renaissance. International Conference-The future of Education." 17 June 2011. Web. 25 October 2015.

Rebhorn, Wayne A. *Courtly Performances: Masking and Festivity in Castiglione's Book of the Courtier*. Detroit: Wayne State University Press. 1978. Print.

Rickaby, John. "Cardinal Virtues." *Catholic Encyclopedia*. New York: Robert Appleton Company. 1913. Web. 25 October 2015.

Roberts, James A., Yaya, Luc Honore Petnji, & Manolis, Chris. "The Invisible Addiction: Cell-phone Activities and Addiction among Male and Female College Students." *Journal of Behavioral Addictions* 1.1 (2014): 1. Web. 25 October 2015.

Roth, Michael. "What's a Liberal Arts Education Good For?" *The Huffington Post*. 1 January 2009. Web. 25 October 2015.

Shankman, Steven, & Durrant, Stephen. *The Siren and the Sage: Knowledge and Wisdom in Ancient Greece and China*. London: Cassel. 2000. Print.

Sterbenz, Christina. "The 100 Most Important People In History." *Business Insider*. 20 December 2013. Web. 25 October 2015.

Takao, M., Takahashi, S., & Kitamura, M. "Addictive Personality and Problematic Mobile Phone Use." *Cyberpsychology and Behavior* 12.5 (2009): 501–507.

Thomée, Sara, Härenstam, Annika, & Hagberg, Mats. "Mobile Phone Use and Stress, Sleep Disturbances, and Symptoms of Depression among Young Adults—A Prospective Cohort Study." *BMC Public Health* 11 (2011): 66. Web. 25 October 2015.

Wang, L. R., & Basten, S. "Population, Gender and Welfare State." 2013. The University of Hong Kong, Faculty of Social Sciences. Web. 25 October 2015.

Ranking Web of Universities. "Taiwan." Cybermetrics Lab. July 2015. Web. 25 September 2015.

Yan, W. T. "Highly Educated Students Unemployed Still Need to Attend University" (高學歷高失業 還是要念大學). *Taiwan Lih Pao* 台灣立報. 28 December 2007. Web. 25 September 2015.

Yen, C. F, Tang, T. C., Yen, J. Y., Lin, H. C., Huang, C. F., Liu, S. C., & Ko, C. H. "Symptoms of Problematic Cellular Phone Use, Functional Impairment and Its Association with Depression Among Adolescents in Southern Taiwan." *Journal of Adolescence* 32.4 (2009): 863–873. Web. 25 September 2015.

Index

aborigine. *See* indigene.
Althusser, Louis, 38
Appaih, Kwame Anthony, 4
Applied English, x, xii
Armour, Ellen T., 53
Atwood, Margaret, 46
audit culture, 25
Austen, Jane, 11
Austronesian, ix

Barthes, Roland, 55
Bartra, Roger, 63
Bates, Jonathan, 88
Benjamin, Walter, 4, 23–24
Beowulf, 73–74
Berman, Jessica, 4
Berryman, John, 110
Bishop, Elizabeth, 12, 99–101, 106–110, 113–114
Boal, Augusto, 38
Boling, Ronald, 126
Bonduca, 121–122, 130
Bono, Barbara, 71
Bradford, William, 60
Branson, Richard, 158
Brodsky, Joseph, 17
Buell, Lawrence, 101–102, 107, 110, 113
Bulgakov, Mikhail, x
Burke, Kenneth, 23

Castiglione, Baldassare, 148
Chaucer, Geoffrey, 75–76
Chen, Chao-ming, 69–70, 72, 83
Chen, Yilin, 12
Cheng, Elyssa Y., 12
Chiang Kai-shek, 1
children's literature, 9
Clodd, Edward, 9
Commodore Perry, 18
Constantino, Renato, 29–30, 32, 49
cosmopolitan, 1, 4–13
cram schools, viii, xii-xiii, 8, 82
Cymbeline, 121–132

Deleuze, Gilles, 24, 27, 63
Derrida, Jacque, 6

Ellison, Ralph, 23
ESP (English for Specific Purposes), xii
Estok, Simon C., 88
Evasco, Marjorie, 42
exophony, 17–28

FLES (foreign language in the elementary school), 58

Flickr, 108, 112
foreigners teaching English, 3
Freire, Paulo, 38
Frere, Sheppard Sunderland, 125

Gaggi, Silvio, 104
GEAT (Global Education Association in Taiwan), 69–70
Gilbert, Sandra, 110–110
Globe Theatre, the, 140
Goldblatt, Howard, 6
Google Maps, 108
Guattari, Felix, 24, 27, 63

Hagedorn, Jessica, 39–41
Haseltine, Patricia, vii-viii, 105
Hau, Carolyn, 39–40
Hoffman, Eva, 22
Holinshed, Raphael, 123, 126
Huang Chun-Ming, 6
Huang Hsinya, 10
Huang, Peter I-min, 87
Huggan, Graham, 89
Hughes, Langston, 6
hypermedia. *See* hypertext.
hypertext, 99–117

Ileto, Reynaldo, 31
indigene: in Australia, 93, 96; in Taiwan, ix, 10, 94, 124
Irving, Washington, 61–62
Iser, Wolfgang, 56

James, Erin, 88
James, Henry, 5
Jen, Gish, 58–59
Jose, F. Sionil, 42–43

Kerridge, Richard, 88
Kingston, Maxine Hong, 7, 58–59
Kristeva, Julia, 53
Krummel, John, 24, 26–27
Ku, Emerald, 12

Lacaba, Emmanuel, 32–33
Lacaba, Jose F., 32, 34–35
Landow, George, 104, 106
Lee, Ang, 11
Lee, Brian, 5
Lee Yuan-chen, 7
Leggate, Alexander, 129
Legislative Yuan, the (the Legislative Ministry), 122
Lenore, Miriel, 89–93
Li Po, 59–60
Little Prince, The, 149
Lowell, Robert, 12, 99–100, 110–115
Lu Hsiu-lien, 7
Lumbera, Bienvenido, 32–34

Ma, Sheng-mei, vii-viii, 7
MacArthur, Arthur, 29–30
Magic Realism, 39, 49
Maley, Willy, 126
Marcos, Ferdinand, 32
Martz, William, 111
Matsumoto, Kazuhito, 11–12
McKinley, William, 29
Mei-li Island uprising, the, 6–7
Melville, Herman, xiii
Meng Chen, 134–138
Mills, C. Wright, 38
Milton, John, 79
Mitchell, David, x
Mizumura, Minae, 19–20, 25–26
MOE (Ministry of Education, Taiwan), 80, 83, 133
Montoneri, Bernard, 5, 12–13
MOOCs (Massive Online Open Courses), 133
Morgan, Wendy, 104
MOST (Ministry of Science and Technology, Taiwan), 81–83
Murakami, Haruki, 20–23
Murray, Les, 89, 93–94, 96

Nandao (southern islands), ix
National Taiwan University (Taida), 70–72
Nativist literature, 6
New Criticism, 6
Norton Anthology of English Literature, The, 72–73

Okada, John, 11
Ortiz, Will P. 12

Paine, Thomas, 62–63
Philippine English, 42–46
Pound, Ezra, 59–60
Pratt, Mary Louise, 54
Providence University (Taichung), ix, 10, 147

Queen Elizabeth I, 77

Raffel, Burton, 111
Ralph, Iris, 12
Rancière, Jacques, 5, 13n4
Renaissance, 5, 76, 147–160
Rizal, Jose P., 31
Roy, Arundathi, 47
Rudman, Mark, 111

Salanga, Alfrredo Navarro, 36–38
San Juan, E., 35–36, 39
Shakespeare, William: as *manga*, 133–138; *Cymbeline*, 121–132; *Othello*, 140–141; *Shamlet*, 142; Teaching Shakespeare in Southern Taiwan, 77–78; via MOOC (Massive Online Open Courses), 133–146
Sir Gawain and the Green Knight, 74
"self-forming formlessness," 11, 21–27
Shelley, Percy Bysshe, x
Sison, Jose Ma., 31
Slaymaker, Doug, 22
Sōseki, Natsume, 19–20
St. John de Crevecoeur, J. Hector, 62
Stein, Gertrude, 57
Stevenson, Robert Louis, viii
Sui Sin Far, 60
Sunflower Student Movement, the (Taiwan), 121–122, 124
Swift, Jonathan, 77–78

Taiwan's coral reef, 95
Taiwan's Formosan cypress, 94–95
Taiwan's mangroves, 94–95
Tamkang University, vii, 87, 89
Tan, Amy, 7, 58–59
Tawada, Yoko, 17–18, 21–23, 27
TESOL (Teaching English as a Second Language), 8, 82, 87–88
Thomas, Herbert, 103
Tiffin, Helen, 89
Tuan Yi-Fu, 101
Twain, Mark, 57

Umbrella Movement, the (Hong Kong), 124

Venuti, Lawrence, 13n2
Villa, Jose Garcia, 35–36
virtues, 149–153; adapt Renaissance virtues for the twenty-first century, 155–158; East and West, 152–153; in Benjamin Franklin, 150 ; in Cicero, 149; in Plato, 149; in the Old Testament, 149

Wang, I-Chun, 12
Warriors of the Rainbow: Seediq Bale, 130
Wheeler, Wendy, 90
Whitman, Walt, 55
Wikipedia, 109
Wordsworth, William, 5, 9, 99, 102
Wu, E-chou, 12

Yeats, W. B., x, 2, 35
Yep, Lawrence, 11

About the Editors and Contributors

Yilin Chen is assistant professor in the English department at Providence University, Taiwan. She earned a PhD in drama and theater at Royal Holloway, University of London, and a MFA in Theatre at Columbia University. Her research interests include Shakespearean adaptations in East Asia, gender and sexuality in Shakespeare, and creative approaches to teaching of Shakespeare. She published in several performance and theater journals, such as *Studies in Theatre & Performance*, *Journal of Adaptation in Film and Performance*, and *Performing Ethos*.

Elyssa Y. Cheng received her PhD in English from the State University of New York at Buffalo in 2003. She is associate professor of English in the Department of Western Languages and Literature, National University of Kaohsiung, Taiwan, where she teaches Shakespeare and early British literature. She has published articles (in Chinese and English) on the politics and poetics of labor in early modern English drama. Her current research focuses upon the relationship between drama and consumption culture in early modern city comedies.

Patricia Haseltine has a doctorate in folklore from Indiana University and studied Native American literature at the University of Arizona. With an interest in Chinese culture, she has also studied the Chinese language. She has taught English at Tamkang University and was a professor in the English Department at Providence University in Taiwan, the Republic of China, for over thirty years. Her publications in Taiwan include *An Aesthetic Negotiation: The Testimonial Poetry of Joy Harjo, Linda Hogan, Simon Ortiz, and Luci Tapahonso* (2006) and *How Traditional Images Still Speak: Transcultural Interpretations of Folklore in Literature* (2012). Currently her research focuses upon interpretation and reception of indigenous film in Asia.

Emerald C. L. Ku is associate professor in the Department of Foreign Languages and Literature, at Asia University. Her recent publications include "Reinscribing the Sense of Place: Elizabeth Bishop's Topography of New England" (2014) and "Tribal Identity and Repositionalization of Multicultural Identity: On Native American Narrative Poetry" (2009). Her research focuses on African American studies, Native American studies, cultural studies, gender studies, and feminism.

Sheng-mei Ma is professor of English at Michigan State University in Michigan, U.S.A., specializing in Asian Diaspora/Asian American studies and East-West comparative studies. His seven single-authored, scholarly books in English are *The Last Isle: Contemporary Film, Culture and Trauma in Global Taiwan* (2015); *Alienglish: Eastern Diasporas in Anglo-American Tongues* (2014); *Asian Diaspora and East-West Modernity* (2012); *Diaspora Literature and Visual Culture: Asia in Flight* (2011); *East-West Montage: Reflections on Asian Bodies in Diaspora* (2007); *The Deathly Embrace: Orientalism and Asian American Identity* (2000); and *Immigrant Subjectivities in Asian American and Asian Diaspora Literatures* (1998). He co-edited and translated *Chenmo de shanhen* (Silent Scars: History of Sexual Slavery by the Japanese Military—A Pictorial Book, bilingual edition, 2005) as well as co-edited *The City and the Ocean* (2011). *Sanshi zuoyou* (Thirty, Left and Right) is his Chinese poetry collection (1989).

Kazuhito Matsumoto is professor of the department of English at Meiji-Gakuin University, Tokyo. He has co-authored *Cutting across Hemingway: Transfigurations of Texts* (in Japanese, 1999), *The Politics of Memory: Forgetting and Remembering in American Literature* (in Japanese, 2001), *The Outline of American Culture* (in Japanese, 2002), and *Transforming the Land: Cities in American Culture* (in Japanese, 2011). He has also co-translated: Ralph Ellison, *Flying Home and Other Stories* (2005), *Shadow and Act* (2009), and with Donald Hall ed., *The Oxford Book of Children's Verse in America* (2008). He is currently working on the project of the "impossibility of the epic in American literature," featuring such writers as William Faulkner and Tim O'Brien.

Bernard Montoneri earned his PhD (African, Arab, and Asian Words; history, languages, literature) and his BA in Chinese from the University of Provence, Aix-Marseille I, France. He is currently associate professor at Tamkang University, Tamsui, Taiwan. He teaches literature (European, children, and British) and languages (French, English, and Italian). Bernard has published about thirty journal and conference papers, five books, and has obtained sixteen teaching or research projects. His research interests include European literature, children's literature, networking, Buddhism, English writing, automated scoring systems, teaching and learning evaluation, data envelopment analysis, and teaching methods. He is the associate editor of *IAFOR Journal of Education*, an invited speaker, and a reviewer for many renowned SCI, SSCI journals such as *Review of Educational Research, American*

Educational Research Journal, *Teaching and Teacher Education*, and *European Journal of Operational Research*.

Will P. Ortiz is assistant professor of Philippine Literature and Creative Writing in Filipino from the University of the Philippines. Her book *Bugtong ng Buwan at Iba Pang Kuwento* (*The Moon's Riddle and Other Stories)* won the Madrigal-Gonzalez Best First Book Award (2012) of the University Press Institute of Creative Writing. She is associate editor of *Daluyan*, journal of the University Press Sentro ng Wikang Filipino and co-editor of *Hardin ng Haraya* (2013), a book of poetry for children. Having a passion for folklore and fantasy literature, she's currently writing a young adult novel, *Ang Pagbabalik ni Maria Makiling*. She awaits the publication of her book *Ang Tatlong Bubwit at ang Bangkang Marikit* (*The Three Little Mice and the Brilliant Boat*), a bilingual picture book about death.

Iris Ralph is assistant professor in the English department of Tamkang University. Between 2006 and 2009, she taught English language and literature at various institutes in Melbourne, Australia, and Incheon, Republic of Korea. Between 1998 and 2005, she taught in the English department of the University of Texas at Austin, U.S.A., where she earned her PhD in English (2005). She has published articles in *NTU Studies in Language and Literature* (National Taiwan University), *Concentric* (National Taiwan Normal University), and *The Ecohumanist* (ASLE-Taiwan). She also has a book chapter published in *Ted Hughes* (2013) and a co-authored book chapter published in *International Perspectives in Ecofeminist Ecocriticism* (2013).

I-Chun Wang teaches English at the Center for Languages and Culture at Kaohsiung Medical University. Since 2014, she holds a joint appointment with National Sun Yat-sen University, where she was formerly professor of English and dean of the College of Liberal Arts and taught Renaissance and twentieth-century dramas, and from 2006 through 2013 directed the Center for the Humanities. Her research interests include comparative literature, and English Renaissance drama. She edited and co-edited books and special issues for journals such as *CLCWEB*, *Cultura: International Journal of Philosophy and Axiology*. Her full-length studies include *Disciplining Women: the Punishment of Female Transgressors in English Renaissance Drama* (1997) and *Empire and Ethnicity: Empire and Ethnic Imagination in Early Modern English Drama* (2011).

E-chou Wu took his PhD in American Literature from Tamkang University, Taipei. He is professor of Literature at Providence University, Taichung, Taiwan. After translating John Updike's *Rabbit Is Rich* into Chinese, published in 2010, he began to take interest in translation studies. He later with full scholarship conducted his second PhD dissertation on adaptation of Ang Lee's films via translation theory at Lingnan University's Department of Translation, Hong Kong, between 2010 and 2013. His past publication is mainly on minority discourse in American literature and intercultural communication.